Dummy variables in econometrics

Econometrics deals with two different types of variables. Firstly, use is made of quantitative variables which represent measurable economic magnitudes, such as income, unemployment, imports, etc. Secondly, quantitative variables are used to indicate attributes, e.g. sex, educational level, race, etc. These two types of variables are referred to as 'real' and 'dummy' variables, respectively. The use in econometrics of dummy variables in the above sense is the topic of this study. Within the class of dummy variables, the binary of zero-one variables will be examined carefully.

The main topics are: the application of binary variables in the linear model; the relations between regression analysis using binary variables as regressors and analysis of variance and covariance; an alternative approach to testing the equality of sets of regression coefficients, applied to the Dutch econometric model 63-D.

Tilburg Studies on Economics 14

Edited by the Tilburg Institute of Economics of the Tilburg School of Economics, Social Sciences and Law

A study on Econometrics

Dummy variables in econometrics

M. A. Kooyman

Department of Econometrics,
University of Groningen

Tilburg University Press
The Netherlands / 1976

Printed in The Netherlands

ISBN 90 237 2919 6

Foreword

Dr. Kooyman has made a careful study of the econometric model used around 1963 by the Netherlands Central Planning Bureau, especially of eight of its most important behavioural equations. The Bureau used coefficients estimated from a pre-war and a post-war time series simultaneously, in order to have a larger number of observations. The author, being aware of the difference in general economic 'climate' in the two periods, tests the assumption of no difference in pre-war and post-war coefficients. In an inventive way he uses a few alternative approaches. On the basis of his findings he concludes that a new specification of some of the equations is necessary. Since the Dutch model is likely to be the oldest of its type, foreign readers may be interested in reading Dr. Kooyman's text - slightly adapted - in English.

J. Tinbergen

Preface

The first edition of this book appeared in Dutch as a thesis which was defended at the State University of Groningen in 1971. When working on this publication I received much help from Dr. H. Rijken van Olst and Dr. G.F.W.M. Pikkemaat, professors at the State University of Groningen of, respectively, statistics and econometrics, and mathematical economics.
I would like to thank Prof.dr. J. Tinbergen, who aroused my enthusiasm for quantitative economics during my college years at the Netherlands School of Economics; I am much obliged to him for his willingness to write a foreword.
To Drs. J.W. Nieuwenhuis and Drs. W. Voorhoeve, from the Econometric Institute of the State University of Groningen, I am obliged for the calculations they have performed. I want to thank Drs. G. Joosten, also of the Econometric Institute, for his helpful criticism after the Dutch edition was published. The amount of typing inherent in the birth of a book was taken care of by Mrs. K. ten Cate-Schuil; I have nothing but praise for her devotion. Mrs. A.C.A. Elderson de Boer did the translation; I think she has done a difficult job very well.
As far as the symbols in this study are concerned, the following remarks must be made. Vectors and matrices have been made recognizable by underlining them. Estimated quantities are as a rule indicated by a $\wedge$ over the symbol in question. In the footnotes the indication 'ibid.' will frequently be found followed by two references; in those

cases the first reference indicates the place in this volume where the publication referred to is mentioned in full, the second serves to indicate the precise place in the original.

M. A. K.

Contents

1. Introduction and formulation of the problem

In econometric research and econometric analysis, 'real' and 'dummy' variables are made use of. Until now, the methods applicable to working with dummy variables have been developed and elaborated less than those for 'real' variables. Little research has as yet been carried out into the possibilities dummy variables offer in analyses. The object of the present study is to demonstrate by means of some elaborated examples that dummy variables are a much more powerful tool in econometric analysis than is evident from present literature.
First the character of real and dummy variables will be explained.
A variable the observations of which can be determined by 'real' measurements of the (economic) magnitude concerned will be called a real variable. Such measurements have been defined by means of a scale which has a logical zero point, and on which the values 0, 1, 2, 3, etc. lie at equal distances from each other, which all have the same meaning (ratio scales). Most economic variables belong to the category of real variables; volume of production, imports and exports, level of income, level of prices, etc. Besides magnitudes that are susceptible of real measuring, there are many whose 'level' can be represented only by code numbers. For such magnitudes, the zero point is not logically fixed but 'chosen'; the intervals 0-1, 1-2, etc. do not have the same meaning. These magnitudes will be referred to as

'qualitative magnitudes'[1]; an example of such a magnitude is a consumer's appreciation of the quality of a product. Qualitative magnitudes sometimes play an important part in economics. The presence or absence of major strikes, to give an example, may be a significant factor in the quantification of an economic model. In order to use this factor for estimating the parameters of the model, one must convert it into numerical observations. For that purpose any pair of unequal code numbers can be chosen, namely one value to indicate the occurrence of major strikes and another to indicate their absence. In many cases the values 0 and 1 are used, e.g. 1 for the occurrence and 0 for the absence of major strikes. Variables representing qualitative magnitudes, as in the example given, will henceforth be referred to as dummy variables[2].

Dummy variables that can adopt only two values - as in the above example - will henceforth be called 'binary dummy variables'. Watts[3] gives a few other names for this category of dummy variables, such as zero-one variables and Kronecker variables.

Although binary dummy variables will be used in the

1. In G.U. Yule and M.G. Kendall (An introduction to the theory of statistics, London 1965, p. 1 ff.) the difference between magnitudes susceptible of real measuring and qualitative magnitudes is discussed extensively. They indicate the difference by the names 'variables' and 'attributes'.
2. In the Dictionary of statistical terms, by M.G. Kendall and W.R. Buckland, under 'dummy variable' not only this type of dummy variable is mentioned, but also the kind of variable consisting of observations all equal to 1. They even prefer using the term 'dummy variable' exclusively in that case. In the present study the term 'dummy variable' will emcompass a wider range of variables than is found with Kendall and Buckland.
3. H.W. Watts, An introduction to the theory of binary variables (or all about dummies), Systems formulation and methodology workshop paper 6406, 1964, Wisconsin University.

majority of examples to be discussed, there are other kinds of dummy variables. A 'variable' with a value 1 for every observation is sometimes also called a dummy variable[4], but where there is no variation one can hardly speak of a variable. Rather we shall refer to this type of dummy variables as 'constant-term variables', because they serve as such in econometrics. There are also dummy variables that are composed of more than two different values. For the quantity 'level of education', e.g., the levels lower, medium and higher can be distinguished. 'Lower level of education' could be denoted by the value 0, 'medium' by 1, and 'higher' by 2. As in the former cases cited, the numerical values used do not represent the results of real measurements; they are, in fact, code numbers. In this study dummy variables built up of more than two different values will be indicated as 'multi-valued dummy variables'.

In econometrics, still other variables are known which no more than dummy variables consist of observations resulting from real measurements. To this category of variables belong, e.g., principal components, and the factors in factor analysis. Unlike dummy variables, however, both last-mentioned variables can be derived from real variables and could, therefore, be called 'derived variables'. The term 'dummy variables' will in this study be reserved exclusively for binary dummy variables, multivalued dummy variables and constant-term variables.

When using multivalued dummy variables in combination with real variables, one should be aware of the difference in character between the numerical values representing the observations of the two kinds of variables. It would be incorrect, for example, to conclude from the

4. See, e.g., E. Malinvaud, Statistical methods of econometrics, Amsterdam 1966, p. 241.

equality of the difference between the values 0, 1 and 2, that the differences in level of education have the same meaning. Code numbers do not work that way. The qualitative magnitude 'level of education' could be represented just as well by another series of code numbers, say 0, 2 and 5, in which the intervals are unequal. It will become clear later on that the different character of the numerical values of the observations, together with the great arbitrariness in the choice of code numbers, are not conducive to a correct application of multivalued dummy variables in econometrics. A serious objection, directly related to the wide freedom of choice of the code numbers to be operated, is that in a linear relation with a multivalued dummy variables as regressor, the regression coefficients of the real variables will be affected by the code numbers chosen; this will be demonstrated in Chapter 2. Binary dummy variables do not show this effect. In Chapter 2 it will also be shown that multivalued dummy variables are, in fact, unnecessary, because any qualitative magnitude can be represented by one or more binary dummy variables.

A survey will be given now of the various ways dummy variables have been used in literature. Next, the objectives of this study will be exposed, after which a short summary of the contents of the various chapters will be given.

Dummy variables as described on page 2 are used to express the influence of qualitative magnitudes in relations between economic magnitudes. This can be done in two distinct ways and for two distinct purposes, viz.;

A. Dummy variables acting in the same way as real variables, notably as regressands or regressors in linear relations;

B. Dummy variables applied to express how a qualitative magnitude causes discontinuous changes in the level of the parameters of the real variables in a linear relation. In this case functions of real and dummy variables act as regressors.

Both usages will now be treated more in detail.

A. Dummy variables as regressors or regressands[5]

First, the use of binary dummy variables will be demonstrated, and next, that of multivalued dummy variables.

Lee[6] 'explains' the price of a house purchased, among other factors, by a number of characteristics of the buyer. Binary dummy variables are introduced as explanatory variables for such characteristics as married or non-married, white or non-white, high school finished or not, resident of a southern state or not, and for a large number of similar characteristics. For individuals possessing the characteristic, the relative binary dummy variable is given the value 1; for individuals lacking it, the value is 0. Another example is found with Hervey[7], who tries to 'explain' differences between median incomes in the U.S.A. from differences in age, schooling and residential area of the people concerned.

---

5. No attention will be given in this review to the constant-term variable; its function is sufficiently known.
6. Lee, Tong Hun, Demand for housing: a cross-section analysis, Review of Economics and Statistics, Vol. 45 (1963), pp. 190-196.
7. J.L. Hervey, A regional analysis of the effects of age, education and occupation on median income, Journal of Regional Science, Vol. 6 (1966), no. 2, pp. 35-48.

Neter & Maynes[8] undertook an investigation into the use of binary dummy variables as regressands in articles published in Volumes 1955-1966 of the Journal of the American Statistical Association and the Review of Economics and Statistics. Only in nine cases did they find a binary dummy variable used, for instance as a variable indicating the purchase or non-purchase of a new car in a certain year.

Tintner[9] also uses a binary dummy variable as a regressand in so-called discriminant analysis.

In Menges[10] binary dummy variables are found used in a totally different situation. He expresses the influence of the reunification of Western Germany with the Saar and West Berlin in 1960 by introducing into his model for Western Germany a so-called 'Springvariable' as a regressor, giving it the value 0 for the years 1951-1959 and the value 1 for the years 1960-1963.

In their Japanese model, Klein and Shinkai[11] use a binary dummy variable in a similar way as Menges; to indicate differences between pre-war and post-war economic structures they add binary dummy variables to 9 out of 15 behavioural equations.

In the Brookings model[12], 18 binary dummy variables are found among the regressors, used for widely varying

---

8. J. Neter and E.S. Maynes, On the appropriateness of the correlation coefficient with a 0,1 dependent variable, Proceedings of the Business and Economic Statistics Section of the American Statistical Association, 1967, pp. 447-453.
9. G. Tintner, Econometrics, New York, 1952, p. 97.
10. G. Menges, Ökonometrische Prognosen, Cologne and Opladen, 1967, p. 32.
11. L.R. Klein and Y. Shinkai, An econometric model of Japan 1930-1959, International Economic Review, Vol. 4 (1961), no. 1, pp. 1-28.
12. J.S. Duesenberry, G. Fromm, L.R. Klein and E. Kuh (eds.), The Brookings quarterly econometric model of the United States, Amsterdam 1965, pp. 745-746.

phenomena.
The last example of this survey, which in no way pretends to be complete, is the application by Lüdeke[13] of a binary dummy variable as a regressor indicating changes in the system of social insurances.
There remain two important uses of binary dummy variables to be mentioned under heading A, viz. their application in time-series analysis and in variance-covariance analysis.
Klein and others[14] add to the regressors of each behavioural equation in their quarterly model three so-called 'quarterly variables'. These three binary dummy variables, denoted as Q_{1t}, Q_{2t} and Q_{3t}, are built up as follows. Variable Q_{1t} takes the value 1 in the first quarter and the value 0 in the other quarters; the other binary dummy variables vary correspondingly. Klein and others use these variables in every behavioural equation containing uncorrected quarterly observations as real variables, to show the influence of seasonal fluctuations. In section 2.5 of this study their method will be reverted to.
A number of authors have discussed the similarity between (1) regression analysis with binary dummy variables as regressors and analysis of variance, and (2) between regression analysis with both real and dummy regressors and analysis of covariance. The former similarity has been analysed by, among others, Malinvaud[15], Goldberger[16],

---

13. D. Lüdeke, Ein ökonometrisches Viertaljahresmodel für die Bundesrepublik Deutschland, Tübingen 1969, p. 63.
14. L.R. Klein, R.J. Ball, A. Hazlewood and P. Vandome, An econometric model of the United Kingdom, Oxford 1961, pp. 40 ff.
15. Ibid., p. 3, pp. 228 ff.
16. A.S. Goldberger, Econometric theory, New York 1964, pp. 227 ff.

and Kendall and Stuart[17]. Graybill[18] hints in the direction of the latter similarity in this discussion of mixed models, while Johnson[19] deals with the matter thoroughly. Linear relations between dummy variables only are disregarded because of the almost complete lack of applications in econometrics[20].

The above list of situations in which binary dummy variables are used may illustrate sufficiently the varied applicability of these variables as regressors and regressands.

Next, some cases of the application of multivalued dummy variables will be presented. Compared with binary dummy variables multivalued dummy variables occur only sporadically. In the Brookings model follow-up study[21] two cases are cited however. First, there is a variable DMY 54, assuming the values 3, 2, 1 and 0 according to the scarcity of motorcars in the U.S.A. right after the Second World War. Second, a variable DMY 53 with values 0, ½ and 1, is applied to indicate the degree of flexibility of prevailing hire-purchase conditions.

In this context it seems fitting to point out how the Central Planning Bureau (C.P.B.) in the Netherlands in

---

17. M.G. Kendall and A. Stuart, The advanced theory of statistics, Vol. 3, London 1966, pp. 6 ff.
18. F.A. Graybill, An introduction to linear statistical models, Vol. 1, New York 1961, chapter 18, Mixed models.
19. J. Jonhston, Econometric Methods 2nd edition, New York 1972.
20. The topic of regressions between dummy variables only is discussed in H.T. Amundsen, Binary variable multiple regressions, Scandinavian Journal of Statistics, Vol. 1 (1974), pp. 59-70.
21. J.S. Duesenberry, G. Fromm, L.R. Klein and E. Kuh (eds.), The Brookings model; some further results, Amsterdam 1969, pp. 81-82, resp.

one of its models[22] uses the variables k and k' to indicate the degree of restriction vs. liberalization of importation into the Netherlands. Variable k can adopt the values 0, 1, 2 and 4, k' the values 0, 1, 3, 4, 10 and 14. The C.P.B. characterizes these variables as so-called 'cleaning variables'. They suggest that a certain meaning should be attached to the numerals used; thus k and k' would not be dummy variables in the sense of page 2.
After this enumeration of the principal applications of dummy variables as regressors or regressands, the second way if using dummy variables will now be discussed.

B. Dummy variables as a means to help quantify changes in the parameters of real variables.

A brief explanation of the use of dummy variables - in this case binary dummy variables - to indicate discontinuous changes in the parameters of a linear relation between real variables, is found in Johnston[23] and Goldberger[24]. Johnston gives as an example a consumption function to be estimated with the help of observations in years of war and peace. Provided there is not a priori condition that the parameters of this function are equal in both periods, the situation can be

---

22. The model referred to, 63-D, has not appeared in any official C.P.B. publication, but in the following publications:
    a. P.J. Verdoorn and J.J. Post, Short and long term extrapolations with the Dutch forecasting model 63-D, discussion note at the meeting in May 1964 of the Centre international d'étude des problèmes humains, Monaco, and b. Macro-economic models for planning and policy-making, publication of the U.N., Geneva 1967, pp. 35-51.
23. Ibid., p. 8, p. 177.
24. Ibid., p. 7, p. 225.

represented as follows:

$$C_t = \alpha_1 + \beta_1 Y_t + \varepsilon_t \quad \text{for war years}$$
$$C_t = \alpha_2 + \beta_2 Y_t + \varepsilon_t \quad \text{for peace years} \qquad (1.1.1)$$

in which

C_t represents the consumptive expenses in period t;

Y_t is the disposable income in period t;

$\alpha_1, \beta_1, \alpha_2$ and β_2 are the parameters, and

ε_t is the disturbance term of the equation in period t.

With the help of a binary dummy variable Z_t, with the value 1 for war years and 0 for peace years, Johnston writes the two separate relations of (1.1.1) as one relation, as follows:

$$C_t = \alpha_2 + (\alpha_1 - \alpha_2) Z_t + \beta_2 Y_t + (\beta_1 - \beta_2) Z_t Y_t + \varepsilon_t \qquad (1.1.2)$$

It will be noticed that in (1.1.2) the variable Z_t occurs not only as a simple regressor but also as a regressor in the product term $Z_t Y_t$. The occurrence of regressors constituting the product of real variables and binary dummy variables is a phenomenon that is characteristic for cases where binary dummy variables are used to quantify discountinuous changes in the parameters of a relation between real variables. Accordingly, in (1.1.2) the parameter of $Z_t Y_t$ equals the difference between the corresponding parameters in the original relations, viz. β_1 and β_2.

In analysis of covariance a similar situation as the one sketched in (1.1.1) is studied. The analysis of relation

(1.1.2) by familiar regression techniques is closely related to analysis of covariance.

After the above survey of applications, and of possibilities of applying dummy variables, the time has come to present the objectives of the present study.

The situations in which binary dummy variables in linear relations are used in practice have given rise to divergent interpretations of the meaning of the coefficients estimated. It seems useful, therefore, to analyse the meaning of these coefficients. That will be the first objective.

Although the relations between analysis of variance on the one hand and regression analysis on the other, and the relation between analysis of covariance on the one hand and regression analysis with real and dummy regressors on the other, have been indicated in literature, the references have mostly been fragmentary and hardly inspiring to an econometrist. It will be the second objective of this study, therefore, to design a box of tools with which relations can be established between the much-used types of analysis of variance and of covariance and regression of analysis.

Finally, this study aims at analysing in theory, and making operational, the use of binary dummy variables for the quantification of discontinuous changes in parameters. The formulation given here implies that practical application is to be considered part of the objective.

In elaborating the first objective a result will be achieved with regard to procedures for the estimation of parameters in relations with errors in variables. It will be demonstrated (see section 2.6) that Wald's and Bartlett's well-known estimators represent special cases of the use of binary dummy variables as instrumental variables.

It will also become evident that the method to be developed towards the third objective, will enable Chow's test[25] to be generalized, and may thus lead to many new applications.

The above mentioned objectives will be treated as follows in the chapters 2 through 5.

In chapter 2 the meaning of regression coefficients in linear relations including dummy variables will be examined. With respect to multivalued dummy variables it will be shown that regression coefficients are sensitive to the values attached to the dummy variables used. For the case of binary dummy variables it will be proved that the regression coefficients of the real variables in the relation studied are invariant as far as the values of the binary dummy variables are concerned.

The second objective will be dealt with in chapter 3. Three frequently used types of analysis of variance and one type of analysis of covariance will be studied from the point of view of regression analysis, viz. the one-way analysis of variance, the two-way analysis of variance with and without interactions and a simple analysis of covariance with one co-variable.

In chapter 4 the theory of the use of dummy variables for quantifying discontinuous changes in the parameters of real variables is formulated; in chapter 5 an application is presented. In formulating the theory much attention will be given to a well-known test of Chow, viz. that for equality of corresponding parameters in two linear regressions. This test will be derived in an alternative way, namely by using a binary dummy variable in the way explained on p. 10. The relation between that technique

---

25. G.C. Chow, Tests of equality between sets of coefficients in two linear regressions, Econometrica, Vol. 28 (1960), no. 3, pp. 591-605.

and analysis of covariance will be investigated.
To apply the results of chapter 4, the parameters of the C.P.B. model 63-D will be examined. The C.P.B. has estimated the parameters of each regressor in the model on the basis of observations made both in pre- and in post-war years, an estimating method which assumes that pre- and post-war parameters are equal. Our examination will focus on the question whether or not the coefficients underwent a significant change at the transition from the pre-war to the post-war period, and whether or not the assumption made by the C.P.B. may be considered acceptable.

2. Dummy variables and the linear regression model

2.1. INTRODUCTION

In chapter 1 it has been suggested that operating dummy variables and interpreting the results of regressions containing dummy variables is not a trivial matter. In the present chapter each of the more important applications of dummy variables in econometrics will be dealt with. As the linear regression model is a well-known and much-used estimating model in econometrics, the applications of dummy variables to be discussed will nearly all be within the setting of the linear regression model.
A simple case in which dummy variables are frequently used is that of a linear model with a constant-term variable and one binary dummy variable among the regressors. Examples are to be found in the Brookings model[1], the Japanese model[2], the German models[3] and the World Tin Market model[4]. Sporadically a multivalued dummy variable is used instead of a binary dummy variable in a regression, along with the constant-term variable and real variables; in the follow-up study about the Brookings

1. Ibid., p. 6, ch. 18, the variables denoted by DMY and a consecutive number.
2. Ibid., p. 6, p. 9.
3. See G. Menges, ibid. p. 6, p. 32, and D. Lüdeke, ibid. p. 7, p. 63.
4. M. Desai, An econometric model of the world tin economy, Econometrica, Vol. 34 (1966), no. 1, pp. 111 and 113.

model an example is given[5].

In section 2.2 the introduction of two dummy variables - among which the constant-term variable - into a regression is studied in detail. First, attention is given to regressions of one real variable on two dummy variables alone; the results of this analysis are then used in the examination of the regressions of a real variable on a number of other real variables completed by two dummy variables. Here, the attention is focussed on the meaning of the regression coefficients of dummy variables in the latter kind of regression.

Section 2.2 is concluded with a special case, in which the binary dummy variable once equals one and otherwise consists of zeros, a situation that occurs in Menges's[6] model for Western Germany.

Section 2.3 discusses the same kind of regressions as does section 2.2 but with several binary dummy variables among the regressors beside the constant-term variable. Examples of such regressions are found, e.g., with Lee[7], in the Brookings model[8], and in the English quarterly model[9]. Contrary to the approach of section 2.2, the possibility of including several multivalued dummy variables in a regression is not discussed in 2.3, for one thing because in section 2.2 it has been proved that the use of even one multivalued dummy variable already gives great interpretation difficulties, for another because application of several multivalued dummy variables in one regression has as yet rarely been dealt with in econometric literature.

---

5. Ibid. p. 8, p. 63.
6. Ibid. p. 6, p. 34, the variable ΔZ.
7. Ibid. p. 5.
8. Ibid. p. 6, p. 693 and 695.
9. Ibid. p. 7, p. 249.

A binary dummy variable used as a regressand will be discussed in section 2.4. As has been shown by Goldberger[10] and will be dealt with later on, a binary regressand fits in badly with a linear regression model. Goldberger suggests an alternative in the shape of so-called probit analysis; in section 2.4 that method will not be gone into because it does not make use of dummy variables. What will be discussed is the relation between discriminant analysis and regression analysis with a binary dependent variable.
Section 2.5 contains a discussion of the use of quarterly variables in the type of time series analysis which divides the fluctuations within a time series into a trend component, a seasonal component and an irregular component. A regression with dummy variables among the regressors will be compared with the conventional method of eliminating seasonal fluctutations.
The final section (2.6) of this chapter deals with the application of dummy variables in situations where there are measuring errors in the regressor. Among other things it will be demonstrated that in one special case the use of a binary dummy variable as an instrumental variable yields the well-known Wald-estimator[11].
Linear regression models play an important part in this chapter. Rather than cutting the thread of the argument by deriving properties of the linear regression model, we shall present a survey of some results obtained in linear-model theory; the argumentation can be found in

---

10. Ibid. p. 7, p. 249.
11. A. Wald, The fitting of straight lines if both variables are subject to error. Annals of Mathematical Statistics, Vol. 11 (1940), pp. 284-300.

Goldberger[12], Graybill[13], Theil[14], and Lehman[15]. The economic relation between Λ variables, Y and $X_1, X_2, \ldots, X_{(\Lambda-1)}$, completed by the disturbance term ε, can be written for the tth observation as a linear formula containing Λ unknown parameters $\beta_0, \beta_1, \ldots, \beta_{(\Lambda-1)}$, thus:

$$Y_t = \beta_0 + \beta_1 X_{1t} + \beta_2 X_{2t} + \ldots + \beta_{(\Lambda-1)} X_{(\Lambda-1)t} + \varepsilon_t \tag{2.1.1}$$

The Λ-1 regressors represent a selection of the most important among the many influences working on the variable Y_t: the effect of the neglected regressors is included in the disturbance term ε_t, which, with an appeal to the central limit theorem is supposed to be normally distributed. In the stochastic specification of relation (2.1.1) it is furthermore assumed that the regressors are not stochastic; only the regressand and the disturbance term are supposed to be stochastic. Relation (2.1.1) extends to T observations (t = 1,2,..., T), to be represented as follows:

$$\begin{bmatrix} Y_1 \\ Y_2 \\ \vdots \\ Y_T \end{bmatrix} = \begin{bmatrix} 1 & X_{11} & X_{21} & \cdots & X_{(\Lambda-1)1} \\ 1 & X_{12} & X_{22} & \cdots & X_{(\Lambda-1)2} \\ \vdots & \vdots & \vdots & & \vdots \\ 1 & X_{1T} & X_{2T} & \cdots & X_{(\Lambda-1)T} \end{bmatrix} \begin{bmatrix} \beta_0 \\ \beta_1 \\ \vdots \\ \beta_{(\Lambda-1)} \end{bmatrix} + \begin{bmatrix} \varepsilon_1 \\ \varepsilon_2 \\ \vdots \\ \varepsilon_T \end{bmatrix} \tag{2.1.2}$$

---

12. Ibid. p. 7, chapter 4.
13. Ibid. p. 8, chapter 6.
14. H. Theil, Principles of Econometrics, New York 1971, chapter 3.
15. E.L. Lehman, Testing statistical hypothesis, New York 1959, chapter 7.

For better manageability, system (2.1.2) is written as follows in matrix algebra (vectors and matrices are underlined from now on):

$$\underline{Y} = \underline{X}\underline{\beta} + \underline{\varepsilon} \tag{2.1.3}$$

For the linear relation of (2.1.3) the following assumptions are made[16]:

(i) $\underline{\varepsilon}$ is normally distributed
(ii) $E\varepsilon_t = 0$ $(t = 1,2,\ldots,T)$
(iii) Matrix $\underline{X}$ is of rank Λ; $\underline{X}$ is not stochastic.

The covariance matrix of the disturbances is indicated by $E\varepsilon_s\varepsilon_t = \omega_{st}$ $(s, t = 1,2,\ldots,T)$. Next, concentrating on minimizing a function written as a quadratic from containing matrix $\underline{\Omega} = [\omega_{st}]$, thus:

$$(\underline{Y} - \underline{X}\underline{\beta})' \underline{\Omega}^{-1} (\underline{Y} - \underline{X}\underline{\beta}), \tag{2.1.4}$$

one obtains the so-called generalized least-squares estimator for $\underline{\beta}$, of the following form:

$$\underline{\hat{\beta}} = (\underline{X}' \underline{\Omega}^{-1} \underline{X})^{-1}\underline{X}' \underline{\Omega}^{-1} \underline{Y}. \tag{2.1.5}$$

The estimator $\underline{\hat{\beta}}$ has the attractive quality of minimum variance within the class of linear and unbaised estimators. Moreover, $\underline{\hat{\beta}}$ follows a multidimensional normal distribution with expectation $\underline{\beta}$ and covariance matrix $(\underline{X}' \underline{\Omega}^{-1} \underline{X})^{-1}$.

Frequent use is made in econometrics of a linear regression model, under the assumption that, as far as variances

---

16. The operator E indicates that of the random variable in question the mathematical expectation should be taken.

and covariances of the disturbances are concerned[17]:

$$\text{(iv)} \quad E\varepsilon_s\varepsilon_t = \delta_{st}\sigma^2 \quad (s,\ t = 1,2,\ldots,T).$$

The regressions in such a model are sometimes indicated as regressions with independent and homoscedastic disturbances. Owing to the changed assumption with respect to the variances and covariances of the disturbances, the function to be minimized passes into:

$$(\underline{Y} - \underline{X}\underline{\beta})'(\underline{Y} - \underline{X}\underline{\beta}) \tag{2.1.6}$$

while the relevant least-squares estimator for $\underline{\beta}$ assumes the well-known form:

$$\hat{\underline{\beta}} = (\underline{X}'\underline{X})^{-1}\underline{X}'\underline{Y} \tag{2.1.7}$$

The covariance matrix of $\hat{\underline{\beta}}$ now becomes $\sigma^2(\underline{X}'\underline{X})^{-1}$. This matrix, with on its main diagonal the variances of the elements of $\hat{\underline{\beta}}$ and outside the main diagonal its covariances, contains the unknown parameter σ^2. To test hypotheses about the elements of $\underline{\beta}$ and to determine the confidence intervals of the regression coefficients, an estimator for σ^2 is required. For that purpose the least-squares estimator:

$$\hat{\sigma}^2 = \frac{1}{(T-\Lambda)}\,\hat{\underline{\varepsilon}}'\hat{\underline{\varepsilon}} \qquad (\hat{\underline{\varepsilon}} = \underline{Y} - \underline{X}\hat{\underline{\beta}}) \tag{2.1.8}$$

is used.
The square of multiple correlation coefficient R, indicating to what extent one has succeeded in 'explaining' the variations in the regressand by means of the fluctuations

---

17. The kronecker delta δ_{st} adopts the value one for $s = t$ and zero for $s \neq t$.

of the regressors, can finally be defined as[18]:

$$R^2 = 1 - \frac{\underline{\hat{\varepsilon}}'\underline{\hat{\varepsilon}}}{\underline{Y}'\underline{Y} - \frac{1}{T}\left(\sum_{t=1}^{T} Y_t\right)^2} \tag{2.1.9}$$

Now a number of tests will be discussed relating to hypotheses regarding the elements of the parameter vector $\underline{\beta}$; these tests will be used in the chapters 3 and 4. The linear relation (2.1.3), in accordance with the desired partitioning of $\underline{\beta}$, is written as:

$$\underline{Y} = \underline{X}_1\underline{\beta}_1 + \underline{X}_2\underline{\beta}_2 + \underline{\varepsilon} \tag{2.1.10}$$

If a hypotheses is made about the last r[19] out of the Λ parameters, then the symbols of (2.1.10) have the following meanings:

$\underline{Y}$: vector of the T observations of the regressand
$\underline{X}_1$: matrix with T observations of the first Λ-r regressors
$\underline{\beta}_1$: vector of Λ-r parameters
$\underline{X}_2$: matrix of the T observations of the last r regressors
$\underline{\beta}_2$: vector of r parameters
$\underline{\varepsilon}$: vector of the T disturbances.

The tests on the parameters contained in chapters 3 and 4 are all significance tests, so that only null hypotheses of the form $\underline{\beta}_2 = \underline{0}$ will be examined. The possible cases are thus confined to various sizes of r.

18. See Goldberger, ibid. p. 7, p. 160.
19. By changing the order of the regressors, any r elements of $\underline{\beta}$ can be put in the last r places. The approach followed does not, therefore, limit the application of the test in any way.

Consecutively, the following hypotheses will be examined:

(i) $r = r$ with $1 < r < \Lambda$

(ii) $r = \Lambda-1$; the null hypothesis concerns all parameters except the constant term

(iii) $r = 1$, the null hypotheses consists of one parameter.

(i) As a test statistic pertaining to the null hypothesis $\underline{\beta}_2 = \underline{0}$, Lehman[20] gives

$$Q_1 = \frac{T-\Lambda}{r} \frac{\hat{\underline{\varepsilon}}_0{}'\hat{\underline{\varepsilon}}_0 - \hat{\underline{\varepsilon}}'\hat{\underline{\varepsilon}}}{\hat{\underline{\varepsilon}}'\hat{\underline{\varepsilon}}} \tag{2.1.11}$$

where $\hat{\underline{\varepsilon}}_0$ represents the disturbance from (2.1.10), estimated with least squares under H_0, while $\hat{\underline{\varepsilon}}$ stands for the least-squares estimator of $\underline{\varepsilon}$ from model (2.1.10). The statistic Q_1 follows, under the null hypothesis, the central F-distribution with r degrees of freedom in the numerator and $T-\Lambda$ in the denominator. The critical region is delimited as follows at a level of 5 per cent: reject H_0 if

$$Q_1 > F_\alpha[r, T-\Lambda]$$

in which $F_\alpha[r, T-\Lambda]$ is that value from the F-table beyond which lie exactly 5 per cent of the distribution of a variable distributed according to $F[r, T-\Lambda]$.
As far as the quality of the above mentioned test is concerned Graybill[21] shows that it is a likelihood-ratio test, Lehman[20] proves it to be one of the class of

20. Ibid. p. 17, pp. 270 and 271.
21. Ibid. p. 8, pp. 135 ff.

invariant[22] UMP tests.

To obtain the power of the test, an alternative hypothesis must be made in respect of $\underline{\beta}_2$, for instance:

$$H_1 : \underline{\beta}_2 = \underline{\beta}_2(1)$$

The test statistic Q_1 now follows a non-central F-distribution with non-centrality parameter λ, of which it is true that:[23]

$$\lambda = \frac{1}{2\sigma^2}[\underline{\beta}_2(1)]'[\underline{X}_1'\underline{X}_1 - \underline{X}_1'\underline{X}_2(\underline{X}_2'\underline{X}_2)^{-1}\underline{X}_2'\underline{X}_1][\underline{\beta}_2(1)] \tag{2.1.12}$$

The power of the test can now be determined with the help of the expression

$$p[\beta_2(1)] = \int_{F_\alpha[r,T-\Lambda]}^{\infty} f[r,T-\Lambda,\lambda]\,df \tag{2.1.13}$$

in which the integral must be calculated over the critical region.

(ii) The test statistic pertaining to the null hypothesis for all parameters with the exception of the constant ($r = \Lambda-1$) reads:

$$Q_2 = \frac{T-\Lambda}{\Lambda-1}\,\frac{\underline{Y}'\underline{Y} - \frac{1}{T}\left(\sum_{t=1}^{T} Y_t\right)^2 - \hat{\underline{\varepsilon}}'\hat{\underline{\varepsilon}}}{\hat{\underline{\varepsilon}}'\hat{\underline{\varepsilon}}} \tag{2.1.14}$$

---

22. Invariance here means, among other things, that the observations in $\underline{X}$, $\underline{X}_1$ and $\underline{X}_2$ may all be decreased or increased by the same constant without the rest being attached.
23. See Graybill, ibid. p. 8, p. 138.

the estimated disturbances being, under the null hypothesis, those from the regression of $\underline{Y}$ on the constant-term variable. If (2.1.14) is looked at in relation to the formula for R^2 in (2.1.9), then Q_2 can be rewritten into the familiar[24] statistic distributed according to $F[\Lambda-1, T-\Lambda]$:

$$Q_2 = \frac{T-\Lambda}{\Lambda-1} \frac{R^2}{1-R^2} \tag{2.1.15}$$

The non-centrality parameter from (2.1.12) will also adopt a simpler from, because $\underline{X}_1 = \underline{1}$, $\underline{1}$ representing a column of ones.

(iii) In case one wants to test the significance of only one of the parameters, $(r = 1)$, the test statistic (2.1.11) passes into the following statistic distributed according to $F[1, T-\Lambda]$:

$$Q_3 = (T-\Lambda)\frac{\hat{\underline{\varepsilon}}'_{00}\hat{\underline{\varepsilon}}_{00} - \hat{\underline{\varepsilon}}'\hat{\underline{\varepsilon}}}{\hat{\underline{\varepsilon}}'\hat{\underline{\varepsilon}}} \tag{2.1.16}$$

in which $\hat{\underline{\varepsilon}}_{00}$ represents the least-squares estimator of $\underline{\varepsilon}$ under $H_0 : \underline{\beta}_2 = \beta_\Lambda = 0$. It can be proved[25] with relative ease that the following equation holds:

$$Q_3 = \left(\frac{\hat{\beta}_\Lambda}{s(\hat{\beta}_\Lambda)}\right)^2 \tag{2.1.17}$$

in which

$\hat{\beta}_\Lambda$: the least-squares estimator for the last parameter in $\underline{\beta}$ in (2.1.3):

---

24. See, e.g., G. Tintner, Handbuch der Ökonometrie, Berlin 1960, p. 124.
25. See Goldberger, ibid. p. 7, p. 181.

$s(\hat{\beta}_\Lambda)$: the estimated standard deviation of $\hat{\beta}_\Lambda$, i.e. the root from the last element of the main diagonal of $\hat{\sigma}^2 C$, C being equal to $(X'X)^{-1}$.

It is furthermore true that the square root from Q_3 follows a $t[T-\Lambda]$-distribution. Test statistic (2.1.17) is therefore equivalent with the familiar t-test on the significance of one parameter, which for the ith element of $\underline{\beta}$ reads

$$Q_4 = \frac{(\hat{\beta}_i - 0)}{\hat{\sigma}\sqrt{c_{ii}}} \qquad i = 1,2,\ldots,\Lambda \tag{2.1.18}$$

Calculation of the power of this test can be carried out as explained under (i).

From the foregoing it follows that all the well-known significance tests on the parameters in a linear model can be reduced to the test described under (i). The foundation for an integral approach to the hypotheses to be tested in the coming chapters has thus been laid.

2.2. ONE BINARY OR MULTIVALUED DUMMY VARIABLE AS A REGRESSOR

It follows from the introductory section to this chapter that in econometrics regressions are carried out which include, beside the constant-term variable, a number of real variables - including the regressand - and one binary or multivalued dummy variable, to be represented by the symbol Z. The meaning of the parameters of dummy variables cannot be the same as that of the parameters of real variables. While the latter represents the change the regressand experiences as a consequence of the relevant regressor's changing by one unit, the former cannot do

so, because the code numbers denoting dummy variables do not represent a certain number of measuring units. In this section it will be proved that the regression coefficient of a binary dummy variable Z in the kind of regression mentioned above corresponds with the difference between least-squares-estimators of the constant term from two regressions: first, the regression between the relevant variables without Z for those observations for which Z takes the value 0, and second, the regression between the variables without Z, for which Z takes the value one. For the Japanese model,[26] which contains binary dummy variables whose value is zero for pre-war observation years and one for post-war observation years, it would be necessary in this context to compare the outcome of the regression for the whole period as far as the coefficient of Z is concerned with the estimates of the constant term in the separate regressions of, respectively, the pre-war and post-war periods.
A restriction should be added, however, to the above statement about the meaning of the regression coefficient of the binary dummy variable, namely, that the corresponding parameters for all real variables should be equal for the two regressions, a condition that will be dealt with in detail in the course of the present section. In a case like the one described by Klein and Shinkai, who had at their disposal time series of the Japanese economy from which certain years (mainly war years) were missing, the use of just one binary dummy variable as regressor to express structural change - that is in their view shift in value of the constant term - between the pre-war and the post-war period could achieve the desired result only if all other parameters in the

26. See Klein and Shinkai, ibid. p. 6, p. 9; the dummy variable is called D there.

relation remained unchanged under the transition from pre-war to post-war data.

The assumption of unchanging parameters for the real variables can be expressed by a null hypothesis in the following way. Let the pre-war elements of the time series belonging to a linear relation within the model be denoted by $\underline{Y}_b$ (for the regressand), $\underline{X}_b$ (for the regressors) and $\underline{1}_a$ (for the constant-term variable), and the post-war elements by, respectively, $\underline{Y}_b$, $\underline{X}_b$ and $\underline{1}_b$. The linear relation may then be written as

$$\begin{bmatrix} \underline{Y}_b \\ \underline{Y}_a \end{bmatrix} = \begin{bmatrix} \underline{X}_b \underline{\beta}_b \\ \underline{X}_a \underline{\beta}_a \end{bmatrix} + \begin{bmatrix} \underline{1}_b \beta_{ob} \\ \underline{1}_a \beta_{oa} \end{bmatrix} + \begin{bmatrix} \underline{\varepsilon}_b \\ \underline{\varepsilon}_a \end{bmatrix} \qquad (2.2.1)$$

The pre-war and post-war constants are indicated in (2.2.1) by β_{ob} and β_{oa}, and the remaining parameters by $\underline{\beta}_b$ and $\underline{\beta}_a$. The disturbances are indicated by $\underline{\varepsilon}_b$ and $\underline{\varepsilon}_a$. Klein and Shinkai[27] use a graphical method to check the reasonableness of the hypothesis $\underline{\beta}_b = \underline{\beta}_a$. For each equation in the model they draw the regressand values and the values of each of the regressors in scatter diagrams, and then determine by sight whether the pairs of pre- and post-war observations lie on parallel lines. In the English quarterly model[28] this method was applied for the first time to quarterly data; in scatter diagrams the regressand was compared with each of the regressors and according to the degree of parallelity between the regression lines per quarter it was decided whether or not the null hypothesis of constant parameters was acceptable. Comparison of the regressand with each of the regressors separately is not presented as the most recommendable method, and quite rightly so, for the hypothesis concerns

27. Ibid. p. 6, p. 3.
28. See Klein c.s., Ibid. p. 7, pp. 43 ff.

more than one parameter, therefore simultaneous testing is required. Chow[29] has designed a simultaneous test which can be used in this context; it will be elaborated upon in chapter 4.
To clarify the meaning of the regression coefficient of a binary dummy variable in the regression of a real variable on other real variables, the constant-term variable and a dummy variable, we shall first of all consider the simple case of the regression of a real variable on the constant-term variable and one binary or multivalued dummy variable. This approach has two advantages:

1. It provides an insight into the effect of introducing a binary dummy variable, and
2. it clearly points out the disadvantages of using a multivalued dummy variable, and thus justifies omitting the second step, that of introducing a multivalued dummy variable into a linear relation also containing real variables as regressors.

If the vector of T observations of variable V_1 is denoted by $\underline{V}_1$, the vector of the constant-term variable by $\underline{1}$, and that of the binary dummy variable Z_1 by $\underline{Z}_1$, then the linear relation in terms of the parameters γ_{10} and γ_{11}, with disturbance vector $\underline{\varepsilon}_1$[30], reads:

$$\underline{V}_1 = \underline{1}_1\gamma_{10} + \underline{Z}_1\gamma_{11} + \underline{\varepsilon}_1$$

29. Ibid. p. 11.
30. The first index 1 with the regressand, the parameters and the disturbances serves as a distinction from other similar relations yet to be dealt with.

Written in full this relation has the following shape[32]:

$$\begin{bmatrix} V_{11} \\ V_{12} \\ \cdot \\ \cdot \\ \cdot \\ \cdot \\ \cdot \\ \cdot \\ \cdot \\ \cdot \\ \cdot \\ V_{1T} \end{bmatrix} = \begin{bmatrix} 1 & 0 \\ 1 & 0 \\ \cdot & \cdot \\ \cdot & \cdot \\ \cdot & \cdot \\ \cdot & 0 \\ \cdot & 1 \\ \cdot & 1 \\ \cdot & \cdot \\ \cdot & \cdot \\ \cdot & \cdot \\ 1 & 1 \end{bmatrix} \begin{bmatrix} \gamma_{10} \\ \gamma_{11} \end{bmatrix} + \begin{bmatrix} \varepsilon_{11} \\ \varepsilon_{12} \\ \cdot \\ \cdot \\ \cdot \\ \cdot \\ \cdot \\ \cdot \\ \cdot \\ \cdot \\ \cdot \\ \varepsilon_{1T} \end{bmatrix} \qquad (2.2.2)$$

If the mean of the regressand values belonging to the T_2 observations of $\underline{Z}_1$ that are equal to one is denoted by $\bar{V}_{12}$, and, analogously, the mean of the remaining T_1 regressand values by $\bar{V}_{11}$, then the least-squares estimates of the parameters γ_{10} and γ_{11} will be:

$$\hat{\gamma}_{10} = \bar{V}_{11} \quad \text{and} \quad \hat{\gamma}_{11} = \bar{V}_{12} - \bar{V}_{11} \qquad (2.2.3)$$

Figure 1 (see next page) illustrates the estimated relation. The regression line consists of only two points, viz. the one where $Z_{1t} = 0$ and the one where $Z_{1t} = 1$, with, respectively, co-ordinates $(0, \bar{V}_{11})$ and $(1, \bar{V}_{12})$. As is shown in the figure the slope of the regression line corresponds with the difference between the two group means.

Figure 1 can also be related directly to the principle of the least-squares method. This method, when used to determine a regression line, minimizes the sum of the

---

31. The zeros and ones in the binary variable can be interchanged without effect on the results of the analysis, as can be easily verified. The points in the matrix of regressors mean something only within the column concerned, viz. that the previous figure is repeated.

Fig. 1. Scatter diagram with regression line of the dummy variable Z_{1t} and the real variable V_{1t}

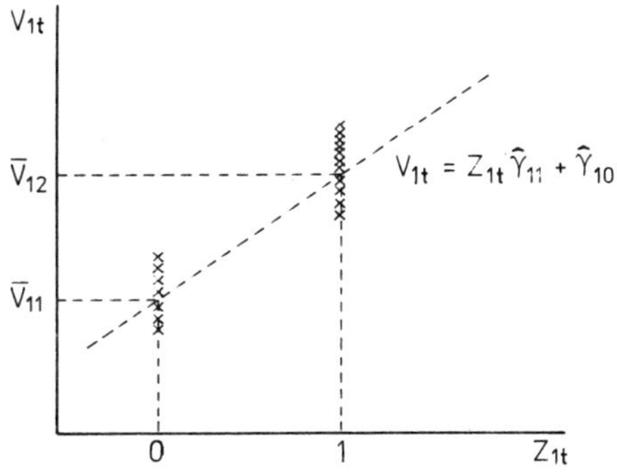

squares of the deviations from the regression line (measured along the vertical axis). Figure 1 shows that to do this, two numbers c_1 and c_2 must be found such that

$$\sum_{t=1}^{T_1} (V_{1t}-c_1)^2 + \sum_{t=T_1+1}^{T} (V_{1t}-c_2)^2$$

is minimized. As is known, that result is achieved for

$$c_1 = \bar{V}_{11} \text{ and } c_2 = \bar{V}_{12}.$$

The estimates for the regressand of (2.2.2) have the following simple structure:

$$\begin{aligned} \hat{V}_{1t} &= \bar{V}_{11}, \text{ for } t = 1,2,\ldots,T_1 \\ \hat{V}_{1t} &= \bar{V}_{12}, \text{ for } t = T_1 + 1,\ T_1 + 2,\ldots,T \end{aligned} \qquad (2.2.4)$$

The estimated disturbances are accordingly:

$$\hat{\varepsilon}_{1t} = V_{1t} - \bar{V}_{11} \quad \text{for } t = 1,2,\ldots,T_1$$
$$\hat{\varepsilon}_{1t} = V_{1t} - \bar{V}_{12} \quad \text{for } t = T_1 + 1,\ T_1 + 2,\ldots,T \tag{2.2.5}$$

As is evident from (2.2.5), the estimated disturbances are equal to the regressand values expressed as deviations from the respective group means. The heterogeneity characterizing the original series of V_{1t} values in respect of their group means is lacking in the estimated disturbances.

Considering the significance test on γ_{11} (see equation (2.1.18) on page 24) in conjuction with (2.2.3), one observes that the same test statistic is obtained as with testing the equality of the means of two normally distributed populations with the same variance σ^2; the latter test is to be found in, among others, Dixon and Massey[32]. Analogously to binary dummy variables, multivalued dummy variables will now be studied, after which regression equations consisting of a real variable as regressand, and real variables, the constant-term variable and a binary dummy variable as regressors, will be dealt with. Multivalued dummy variables are used less frequently than binary dummy variables. In the follow-up study on the Brookings model[33] a multivalued dummy variable is mentioned which takes different values for different levels of a qualitative magnitude, in this case hire-purchase conditions. A multivalued dummy variable may be expressed, e.g., in a number of threes, twos, ones and zeros. Leaving any objections against the translation of

---

32. W.J. Dixon and F.J. Massey, Introduction to statistical analysis, New York 1957, p. 121.
33. Ibid. p. 8.

qualitative magnitude into numerical values apart[34], this section will mainly give attention to the interpretation of regression results obtained with multivalued dummy variables. As with the exposition on binary dummy variables, the starting point will be the regression of a real variable on the constant-term variable and a multivalued dummy variable.

For the argument it suffices to have a three-valued dummy variable Z_2, composed of the elements 0, a and b, with frequencies of, respectively T_1, T_2, and $T_3 = T - T_1 - T_2$. The linear relation between regressand $\underline{V}_2$ and the regressors $\underline{1}_2$ and $\underline{Z}_2$, with parameters γ_{20} and γ_{21} and disturbance vector $\underline{\varepsilon}_2$, reads:

$$\underline{V}_2 = \underline{1}_2\gamma_{20} + \underline{Z}_2\gamma_{21} + \underline{\varepsilon}_2$$

Written in full, this relation reads:

$$\begin{bmatrix} V_{21} \\ V_{22} \\ \cdot \\ \cdot \\ \cdot \\ \cdot \\ \cdot \\ \cdot \\ \cdot \\ \cdot \\ \cdot \\ \cdot \\ \cdot \\ \cdot \\ \cdot \\ V_{2T} \end{bmatrix} = \begin{bmatrix} 1 & 0 \\ 1 & 0 \\ \cdot & \cdot \\ \cdot & \cdot \\ \cdot & \cdot \\ \cdot & 0 \\ \cdot & a \\ \cdot & a \\ \cdot & \cdot \\ \cdot & a \\ \cdot & b \\ \cdot & b \\ \cdot & \cdot \\ \cdot & \cdot \\ \cdot & \cdot \\ 1 & b \end{bmatrix} \begin{bmatrix} \gamma_{20} \\ \gamma_{21} \end{bmatrix} + \begin{bmatrix} \varepsilon_{21} \\ \varepsilon_{22} \\ \cdot \\ \cdot \\ \cdot \\ \cdot \\ \cdot \\ \cdot \\ \cdot \\ \cdot \\ \cdot \\ \cdot \\ \cdot \\ \cdot \\ \cdot \\ \varepsilon_{2T} \end{bmatrix} \qquad (2.2.6)$$

34. See chapter 1, pp. 3-4.

The estimators of γ_{20} and γ_{21} by means of the least-squares method (equation (2.1.7) on page 19) yield expressions in terms of the mutual differences between the group means of the first T_1 observations, the next T_2 observations and the remaining observations, which group means will be indicated by, respectively, $\bar{V}_{21}$, $\bar{V}_{22}$ and $\bar{V}_{23}$. These expressions are:

$$\hat{\gamma}_{20} = \frac{a^2T_1T_2\bar{V}_{21}+b^2T_1T_3\bar{V}_{21}+b(b-a)T_2T_3\bar{V}_{22}+a(a-b)T_2T_3\bar{V}_{23}}{a^2T_1T_2+b^2T_1T_3+(b-a)^2T_2T_3} \qquad (2.2.7a)$$

$$\hat{\gamma}_{21} = \frac{aT_1T_2(\bar{V}_{22}-\bar{V}_{21})+bT_1T_3(\bar{V}_{23}-\bar{V}_{21})+(b-a)T_2T_3(\bar{V}_{23}-\bar{V}_{22})}{a^2T_1T_2+b^2T_1T_3+(b-a)^2T_2T_3} \qquad (2.2.7b)$$

There is little left of the simplicity characterizing the outcomes obtained with binary dummy variables. It does remain true that the slope $\hat{\gamma}_{21}$, like $\hat{\gamma}_{11}$ from (2.2.3), becomes equal to zero if the group means are mutually equal. If the group means show no differences, the constant term in (2.2.3) as well as in (2.2.7a) adopts the value of the arithmetical mean of the regressand. In general, however, there will be differences between the group means, which implies for the regression on a multivalued dummy variable that the complex structure of the regression coefficients in most cases cannot be defended and interpreted on the ground of economic motives.

That situation is mainly due to the circumstance that it is in the nature of the qualitative magnitude represented by the multivalued dummy variable to leave a great number of possible values for a and b to choose from, all causing a different weighting of the group means (see (2.2.7a) and (b)), and so leading to different values of

$\hat{\gamma}_{20}$ and $\hat{\gamma}_{21}$.

Analogously to the procedure followed with binary dummy variables, the estimated disturbances pertaining to (2.2.6) can be calculated. For the V_{2t} values occuring in the second group of T_2 observations, the following estimated disturbances are found

$$\hat{\varepsilon}_{2t} = V_{2t} - \hat{V}_{2t} = V_{2t} - \hat{\gamma}_{20} - a\hat{\gamma}_{21} \doteq$$

$$= V_{2t} - \frac{a^2T_1T_2\bar{V}_{22} + b^2T_1T_3\bar{V}_{21} + (b-a)^2T_2T_3\bar{V}_{22} + abT_1T_3(\bar{V}_{23} - \bar{V}_{21})}{a^2T_1T_2 + b^2T_1T_3 + (b-a)^2T_2T_3} \quad (2.2.8)$$

for $t = T_1 + 1, T_1 + 2, \ldots, T_1 + T_2$.

The dense and complicated formula (2.2.8) bears our that in general it is impossible to express in words the contents of the relation between the regressand values and the estimated disturbances. The rather wide freedom often left by the character of the qualitative magnitude as far as the allotting of numbers to a and b is concerned, and most of all the sensitivity of the regression results that follows from it, give rise to a need for other methods. An alternative approach is available: that of expressing the qualitative magnitude from (2.2.6) by means of two binary dummy variables Z_3 and Z_4, in the following way:

$$\begin{aligned}
Z_{3t} &= 1 \text{ for } t = 1,2,\ldots,T_1 \\
Z_{3t} &= 0 \text{ for } t = T_1 + 1, T_1 + 2,\ldots,T_1 + T_2 \\
Z_{3t} &= 0 \text{ for } t = T_1 + T_2 + 1, T_1 + T_2 + 2,\ldots,T \\
Z_{4t} &= 0 \text{ for } t = 1,2,\ldots,T_1 \\
Z_{4t} &= 1 \text{ for } t = T_1 + 1, T_1 + 2,\ldots,T_1 + T_2 \\
Z_{4t} &= 0 \text{ for } t = T_1 + T_2 + 1, T_1 + T_2 + 2,\ldots,T
\end{aligned} \quad (2.2.9)$$

Anticipating the contents of section 2.3, which will deal with regressions with more than one binary dummy variable, it can be stated here that the regression of $\underline{V}_2$ on the constant-term variable, $\underline{Z}_3$ and $\underline{Z}_4$, yields the following results for the regression coefficients $\hat{\gamma}_{30}$, $\hat{\gamma}_{31}$ and $\hat{\gamma}_{32}$ and the estimated disturbances $\hat{\varepsilon}_{3t}$:

$$\begin{aligned} \hat{\gamma}_{30} &= \bar{V}_{23} \\ \hat{\gamma}_{31} &= \bar{V}_{21} - \bar{V}_{23} \\ \hat{\gamma}_{32} &= \bar{V}_{22} - \bar{V}_{23} \end{aligned} \tag{2.2.10}$$

and

$$\begin{aligned} \hat{\varepsilon}_{3t} &= V_{2t} - \bar{V}_{21} \quad \text{for } t = 1,2,\ldots,T_1 \\ \hat{\varepsilon}_{3t} &= V_{2t} - \bar{V}_{22} \quad \text{for } t = T_1 + 1,\ T_1 + 2,\ldots,T_1 + T_2 \\ \hat{\varepsilon}_{3t} &= V_{2t} - \bar{V}_{23} \quad \text{for } t = T_1 + T_2 + 1,\ T_1 + T_2 = 2,\ldots,T \end{aligned} \tag{2.2.11}$$

In this way the same simple and easy-to-interpret structure has been attained as in the situation with one binary dummy variable. Moreover this method avoids the arbitrariness characterizing the allotment of numbers to the symbols a and b in the multivalued dummy variable.
So much for the introduction preceding the regression of a real variable on real variables completed by the constant-term variable and a binary dummy variable. The results obtained are now ready to be put to use in the remaining part of this section.
It will be demonstrated in the first place that regression of a real variable $\underline{Y}$ on a number of real variables contained in matrix $\underline{X}$, and the constant-term variable and a number of dummy variables contained in matrix $\underline{W}$[35],

---

35. The proof announced here will be given for the general case of p dummy variables. Next, the special case of $\underline{W}$ being composed of the two regressors of (2.2.2) will be dealt with.

corresponds as far as the coefficient of the real variables is concerned with the regression of the estimated disturbances of the regression of $\underline{Y}$ on $\underline{W}$ on the estimated disturbances of the regressions of the variables from $\underline{X}$ on $\underline{W}$[36]. In this connection one should mark that the values of the numbers in the $\underline{W}$-matrix will in general affect the regression coefficients of the real variables from the regression of $\underline{Y}$ on $\underline{W}$ and $\underline{X}$, unless the estimated disturbances from the regressions of $\underline{Y}$ on $\underline{W}$ and of $\underline{X}$ on $\underline{W}$ are invariant for these values; in section 2.3 this subject will again be discussed.
In the second place, the meaning of the regression coefficient belonging to the binary dummy variable from the regression of $\underline{Y}$ on $\underline{X}$ and $\underline{W}$ will be examined, in case $\underline{W}$ consists of the two regressors used in (2.2.2).
In matrix notation the relation between regressand $\underline{Y}$ and regressors $\underline{X}$ and $\underline{W}$ reads:

$$\underline{Y} = \underline{\underline{X}}\underline{\alpha} + \underline{\underline{W}}\underline{\delta} + \underline{\varepsilon} \tag{2.2.12}$$

in which

$\underline{Y}$: regressand vector with T elements
$\underline{X}$: T × (Λ-1) matrix of real variables
$\underline{\alpha}$: parameter vector with (Λ-1) elements α_i $(i = 1,2,\ldots,\Lambda-1)$
$\underline{W}$: T × p matrix of dummy variables, including the constant-term variable

36. See T. Kloek in: Note on convenient matrix notations in multi-variate statistical analysis and in the theory of linear aggregation. International Economic Review, Vol. 2 (1961), no. 3, pp. 351-360.

$\underline{\delta}$: parameter vector with p elements δ_j (j = 1,2,...,p)
$\underline{\varepsilon}$: vector with T disturbances.

The normal equations belonging to the linear model from (2.2.12) read:

$$
\begin{aligned}
\underline{X}'\underline{X}\hat{\underline{\alpha}} + \underline{X}'\underline{W}\hat{\underline{\delta}} &= \underline{X}'\underline{Y} \\
\underline{W}'\underline{X}\hat{\underline{\alpha}} + \underline{W}'\underline{W}\hat{\underline{\delta}} &= \underline{W}'\underline{Y}
\end{aligned}
\qquad (2.2.13)
$$

By pre-multiplying the bottom line from (2.2.13) by $\underline{X}'\underline{W}(\underline{W}'\underline{W})^{-1}$ and deducting it from the top one, $\hat{\underline{\delta}}$ is eliminated, leaving an expression for $\hat{\underline{\alpha}}$:

$$[(\underline{X}'\underline{X}) - \underline{X}'\underline{W}(\underline{W}'\underline{W})^{-1}\underline{W}'\underline{X}]\hat{\underline{\alpha}} = [\underline{X}' - \underline{X}'\underline{W}(\underline{W}'\underline{W})^{-1}\underline{W}']\underline{Y} \qquad (2.2.14)$$

To simplify the expression in (2.2.14), one needs to consider the regressions of the variables from $\underline{X}$ and of the regressand $\underline{Y}$, both on $\underline{W}$. Regression of $\underline{X}$ on $\underline{W}$ yields the matrix of estimated disturbances, to be denoted by $\underline{X}_c$:

$$\underline{X}_c = \underline{X} - \underline{W}(\underline{W}'\underline{W})^{-1}\underline{W}'\underline{X} = [\underline{I} - \underline{W}(\underline{W}'\underline{W})^{-1}\underline{W}']\underline{X} \qquad (2.2.15)$$

If the property that $[\underline{I} - \underline{W}(\underline{W}'\underline{W})^{-1}\underline{W}']$ is an idempotent matrix is made use of, then one may say that:

$$(\underline{X}_c'\underline{X}_c) = [\underline{X}'\underline{X} - \underline{X}'\underline{W}(\underline{W}'\underline{W})^{-1}\underline{W}'\underline{X}] \qquad (2.2.16)$$

In the same way the expression $[\underline{X}' - \underline{X}'\underline{W}(\underline{W}'\underline{W})^{-1}\underline{W}']\underline{Y}$ from (2.2.14) can be written in terms of the estimated disturbances belonging to the regression of $\underline{Y}$ on $\underline{W}$, to be called $\underline{Y}_c$, and those of the regression of $\underline{X}$ on $\underline{W}$:

$$\underline{X}_c'\underline{Y}_c = [\underline{X}' - \underline{X}'\underline{W}(\underline{W}'\underline{W})^{-1}\underline{W}']\underline{Y} \tag{2.2.17}$$

Substitution of the intermediary results from (2.2.16) and (2.2.17) into the formula for $\hat{\underline{\alpha}}$ from (2.2.14) leads to:

$$(\underline{X}_c'\underline{X}_c)\hat{\underline{\alpha}} = \underline{X}_c'\underline{Y}_c \tag{2.2.18}$$

Pre-multiplying (2.2.18) by $(\underline{X}_c'\underline{X}_c)^{-1}$, one gets finally:

$$\hat{\underline{\alpha}} = (\underline{X}_c'\underline{X}_c)^{-1}\underline{X}_c'\underline{Y}_c \tag{2.2.19}$$

The estimator for $\underline{\alpha}$ from (2.2.12) can apparently be obtained by the regression of $\underline{Y}$ on $\underline{X}$ after substituting each variable by the relative estimated disturbances from the regressions on $\underline{W}$.
The estimator for $\underline{\delta}$ can be derived directly from the normal equations of (2.2.13):

$$\hat{\underline{\delta}} = (\underline{W}'\underline{W})^{-1}\underline{W}'\underline{Y} - (\underline{W}'\underline{W})^{-1}\underline{W}'\underline{X}\hat{\underline{\alpha}} \tag{2.2.20}$$

A simpler result is reached by taking instead of $\hat{\underline{\delta}}$ the quantity $\underline{W}\hat{\underline{\delta}}$: from this the estimates for the elements of $\underline{\delta}$ can be derived in a simple way, because matrix $\underline{W}$ is in most cases of a simple structure.

$$\underline{W}\hat{\underline{\delta}} = \underline{W}(\underline{W}'\underline{W})^{-1}\underline{W}'\underline{Y} - \underline{W}(\underline{W}'\underline{W})^{-1}\underline{W}'\underline{X}\hat{\underline{\alpha}}$$

$$= (\underline{Y} - \underline{Y}_c) - (\underline{X} - \underline{X}_c)\hat{\underline{\alpha}} \tag{2.2.21}$$

The above derivations were achieved with a $\underline{W}$-matrix which was not specified. The intention was, however, to examine the coefficient of the binary dummy variable from the regression of $\underline{Y}$ on $\underline{X}$, the constant-term variable and one binary dummy variable. By composing the $\underline{W}$-matrix of the two regressors of (2.2.2), and applying formula (2.2.21), one obtains the following estimators for the two elements of $\underline{\delta}$, viz. δ_0 and δ_1, in which the means of the variables concerned have been called $\bar{Y}_1$ and $\bar{X}_{1i}$ $(i = 1,2,\ldots,\Lambda-1)$ for the first T_1 observations and $\bar{Y}_2$ and $\bar{X}_{2i}$ for the remaining T_2 observations:

$$\begin{aligned} \hat{\delta}_0 &= \bar{Y}_2 - \sum_{i=1}^{\Lambda-1} \bar{X}_{2i}\hat{\alpha}_i \\ \hat{\delta}_1 &= (\bar{Y}_1 - \bar{Y}_2) - \sum_{i=1}^{\Lambda-1} (\bar{X}_{1i} - \bar{X}_{2i})\hat{\alpha}_i \end{aligned} \qquad (2.2.22)$$

As regards the meaning of the coefficient $\hat{\delta}_1$ it is sometimes put forward that it indicates a discontinuous change between the constant terms in two regressions, viz. the regression for those observations for which the binary dummy variable has the value zero, and the regression for the remaining observations. In fact, the following two linear relations are compared, the parameters of which may be different, as the indices a and b indicate:

$$Y_t = \delta_{0a} + X_{1t}\alpha_{1a} + X_{2t}\alpha_{2a} + \ldots + X_{(\Lambda-1)t}\alpha_{(\Lambda-1)a} + \varepsilon_t \qquad (2.2.23)$$

for $(t = 1,2,\ldots,T_1)$, and

$$Y_t = \delta_{0b} + X_{1t}\alpha_{1b} + X_{2t}\alpha_{2b} + \ldots + X_{(\Lambda-1)t}\alpha_{(\Lambda-1)b} + \varepsilon_t$$

for $(t = T_1 + 1, T_1 + 2,\ldots,T)$.

The least-squares estimator for the constants of the above relations are:

$$\hat{\delta}_{0a} = \bar{Y}_1 - \sum_{i=1}^{\Lambda-1} \bar{X}_{1i}\hat{\alpha}_{ia} \quad \text{and} \quad \hat{\delta}_{0b} = \bar{Y}_2 - \sum_{i=1}^{\Lambda-1} \bar{X}_{2i}\hat{\alpha}_{ib} \qquad (2.2.24)$$

Comparing the two coefficients from (2.2.24) with the coefficients of the dummy variables from the regression of $\underline{Y}$ on $\underline{X}$ and $\underline{W}$ (see (2.2.22)) one notices that $\hat{\gamma}_0$ is equal to $\hat{\gamma}_{0b}$ and that $\hat{\gamma}_1$ is equal to $\hat{\gamma}_{0a} - \hat{\gamma}_{0b}$, if the parameters of the real variables may be concerned equal for the two periods. If the equalities $\alpha_{ia} = \alpha_{ib} = \alpha_i$ for $i = 1,2,\ldots,\Lambda-1$ must be rejected on the ground of the observations, then there is no sense in calculating a regression for the whole period. Before interpreting the coefficients from (2.2.22) one should, therefore, test the equality of the parameters of the real variables. In chapter 4 this test will be gone into.

To illustrate the remarks made about including a binary dummy variable in a regression the following example may serve. The macro-economic model 63-D[37] for the Netherlands designed by the Central Planning Bureau of the Netherlands (C.P.B.) contains an private employment equation in which a (private employment) is 'explained' by means of the variables v_a (total expenditure), K (gross profits per unit of output), c^r_{-1} (the deposits at the end of the year), p_{m-v}, (the margin between the price of imports and the price level of total expenditure) and $\Delta\tilde{w}_{\ell,-1}$ (a curvilinear indicator of the use of capacity). The parameters of the model have been estimated with the help of time series covering the periods 1923 up to and including 1938, and 1949 up to and including 1960. The time series of the variables consist of observations in the form of first differences.

The least-squares estimators of the parameters from the

---

37. Ibid. p. 9.

private employment equation will be given for[38]:

A. the whole period
B. the pre-war period
C. the post-war period
D. the whole period, with the addition of a binary dummy variable Z with value zero for pre-war years and value one for post-war years.

A. $$a = \underset{(.05)}{.28v_a} + \underset{(.12)}{.43K} + \underset{(.02)}{.05c^r_{-1}} + \underset{(.02)}{.10p_{m-v'}} - \underset{(.36)}{.88\Delta\tilde{w}_{\ell-1}} + \underset{(.29)}{.49}$$
$$R = .950$$

B. $$a = \underset{(.10)}{.46v_a} + \underset{(.18)}{.38K} + \underset{(.03)}{.01c^r_{-1}} + \underset{(.03)}{.10p_{m-v'}} - \underset{(.69)}{.33\Delta\tilde{w}_{\ell-1}} + \underset{(.44)}{.43}$$
$$R = .972$$

C. $$a = \underset{(.04)}{.28v_a} + \underset{(.42)}{.21K} + \underset{(.04)}{.01c^r_{-1}} + \underset{(.05)}{.08p_{m-v'}} - \underset{(.34)}{.55\Delta\tilde{w}_{\ell-1}} + \underset{(.49)}{.18}$$
$$R = .976$$

D. $$a = \underset{(.05)}{.34v_a} + \underset{(.10)}{.51K} + \underset{(.02)}{.04c^r_{-1}} + \underset{(.02)}{.09p_{m-v'}} - \underset{(.31)}{.71\Delta\tilde{w}_{\ell-1}} + \underset{(.27)}{.83} - \underset{(.40)}{1.21Z}$$
$$R = .966$$

Comparing the regression under B with the one under C, one notices the evident difference between the coefficients of variable v_a. A comparison between the regressions under A and D shows that the regression coefficients hardly differ and that the coefficient for the binary dummy variable is significant on a level of 95 per cent. Trying to interpret the coefficient of

38. The estimated standard deviations are printed (between brackets) below the regression coefficients.

Z - which will in most cases be done without knowing the results of the first three regressions - one will tend to conclude that in the post-war period the positive trend[39] of the pre-war years turn into a negative one. An argument based on economic considerations that would make this conclusion plausible is hard to find. The economic conditions in the two periods would seem to make an inverse conclusion more likely. In chapter 5 it will become evident, in fact, that the first-mentioned conclusion must be discarded as being faulty, for not the constants in the regressions for the pre- and post-war periods show differences, but the parameters of v_a. All this may serve as evidence that indiscriminate addition of a binary dummy variable does not in general contribute towards a meaningful study of the economic structure of a country for separate periods, and that one should rather examine all parameters concerned as to their constancy over the different periods. Chapters 4 and 5 will pursue this idea further.

A special case.
It happens regularly that a binary dummy variable is used to represent unique events, such as the Suez crisis. In such a case a binary dummy variable is used whose elements are all zeros, except one which has the value one. Such a dummy variable can also be obtained by transforming all variables of a regression into first differences; if, e.g., $\underline{Z}$ (see page 40) were among the regressors, then a variable would result containing the number 1 to indicate the step from pre-war to post-war observations and for the rest zeros. An example is to be

39. The variables in model 63-D of the C.P.B. are expressed in first differences and for that reason the constant in an equation is a trend.

found in Menges's[40] model for Western Germany. This type of binary dummy variable will be indicated by $\underline{Z}_e$.
The case we want to present has the form of relation (2.2.12), with the observations of the variables $\underline{Y}$ and $\underline{X}$ consisting of data in, for instance, the shape of first differences, and with $\underline{W}$ composed as[41]:

$$\underline{W} = \begin{bmatrix} 1 & 0 \\ 1 & 0 \\ . & . \\ . & . \\ . & . \\ . & 0 \\ . & 1 \\ . & 0 \\ . & 0 \\ . & . \\ . & . \\ . & . \\ 1 & 0 \end{bmatrix} \qquad (2.2.25)$$

It will be demonstrated that including a $\underline{W}$-matrix of the above content will render the estimated disturbance in the regression of $\underline{Y}$ on $\underline{X}$ and $\underline{W}$ for the kth observation identically equal to zero. This does not imply that the regression coefficient of the binary dummy variable equals the estimated k^{th} disturbance in the regression from which this dummy variable is omitted; the collinearity between the binary dummy variable and the other variables makes itself felt here. As the number of

---

40. Ibid., p. 6, pp. 32 ff.
41. The number 1 figures on the kth place in $\underline{Z}_e$.

observations increases it may be expected, as will become evident below, that the coefficient of the binary dummy variable tends towards the kth estimated disturbance in the corresponding regression without the dummy variable. For the argument it is useful to include the constant-term variable and the real regressors in matrix $\underline{P}$ and to take variable $\underline{Z}_e$ separately. The relation between $\underline{Y}$, the variables in $\underline{P}$, and $\underline{Z}_e$ has the following appearance:

$$\underline{Y} = \underline{P}\underline{\gamma} + \underline{Z}_e\delta + \underline{\varepsilon}_e \qquad (2.2.26)$$

By means of the system of normal-equations

$$\begin{aligned} \underline{P}'\underline{P}\,\hat{\underline{\gamma}} + \underline{P}'\underline{Z}_e\hat{\delta} &= \underline{P}'\underline{Y} \\ \underline{Z}_e'\underline{P}\,\hat{\underline{\gamma}} + \underline{Z}_e'\underline{Z}_e\hat{\delta} &= \underline{Z}_e'\underline{Y} \end{aligned} \qquad (2.2.27)$$

the estimator for δ can be derived simply from the second line of (2.2.27) by filling in $\underline{Z}_e'\underline{Z}_e = 1$:

$$\hat{\delta} = \underline{Z}_e'(\underline{Y} - \underline{P}\hat{\underline{\gamma}}) \qquad (2.2.28)$$

The kth estimated disturbance from the regression of (2.2.26) is calculated from:

$$\hat{\underline{\varepsilon}}_e = Y - \underline{P}\hat{\underline{\gamma}} - \underline{Z}_e\hat{\delta} \qquad (2.2.29)$$

For the kth element of $\hat{\underline{\varepsilon}}_e$ it follows, after filling in $\hat{\delta}$ and if the kth row of $\underline{P}$ is denoted by $\underline{P}_k$:

$$\hat{\varepsilon}_{ek} = Y_k - \underline{P}_k\hat{\underline{\gamma}} - [Y_k - \underline{P}_k\hat{\underline{\gamma}}] = 0 \qquad (2.2.30)$$

since $Z_{ek} = 1$.
As in (2.2.19) the vector of coefficients $\hat{\underline{\gamma}}$ can be express-

ed in the estimated disturbances from the regressions of $\underline{Y}$ and the real variables out of $\underline{P}$ on $\underline{Z}_e$ and the constant-term variable. The relation between the estimated disturbances and the original variables is characterized by the following equations ($\underline{Y}$ serves as an example):

$$Y_{ct} = Y_t - \bar{Y}^* \quad \text{for } t = 1,2,\ldots,k-1,\ k+1,\ k+2,\ldots,T$$
$$Y_{ct} = 0 \qquad\qquad \text{for } t = k \tag{2.2.31}$$

in which $\bar{Y}^* = \frac{1}{T-1} \sum_{\substack{t=o \\ t \neq k}}^{T} Y_t$.

Consequently for large values of T, the effect of the introduction of $\underline{Z}_e$ in a linear equation is the same as omitting the kth observation in the original equation. For a high value of T the influence of omitting one observation from the series is negligible, and the estimator for $\underline{\gamma}$ approaches the estimator for the parameters from the regression of $\underline{Y}$ on $\underline{P}$.

2.3. SEVERAL BINARY DUMMY VARIABLES AS REGRESSORS

The possibility of including in a regression more than one multivalued dummy variable will not be discussed, on the one hand in view of the disadvantages connected with the use of multivalued dummy variables[42], and on the other hand because in standard econometric literature no examples of the occurrence of more than one multivalued dummy variable in one and the same relation can be found. One does find in literature examples of binary dummy variables occurring in regressions at frequencies varying from two or more than ten. In the Brookings model, for

42. See pp. 32 and 33.

instance, one finds in chapter 18[43], in which the model is printed in extenso, three relations with, respectively, two, three and four binary dummy variables; Orcutt et al[44]. constructed a regression with 17 binary dummy variables: Lee[45] comes to 15 binary dummy variables, and in the English quarterly model[46] quarterly variables occur which consist of three binary dummy variables.

The need to operate more than one binary dummy variable makes itself felt when the influence of a qualitative magnitude on the regressand is insufficiently expressed by the distinction of only two states of that variable. Sometimes it is felt that the character of the qualitative magnitude can be done justice to only by introducing a finer subdivision than that of a dichotomy. If the level of schooling of employees is at stake, for instance, a subdivision according to lower, medium and higher can be a necessity within the scope of the regression problem in which the variable is being used.

It is not recommendable to achieve a subdivision by means of a three valued dummy variable composed of, e.g., zeros, ones and twos for, respectively, lower, medium and higher education, for a difference equal to unity between lower and medium level as well as between medium and higher level would imply the equality of these differences in level, and that would be contrary to the qualitative character of the magnitude concerned. Any other three values of the dummy variable would have the same disadvantage of suggesting mutual measurability and quantitative comparability of the differences in level.

---

43. Ibid. p. 6, pp. 693 and 695.
44. G.H. Orcutt, M. Greenberger, J. Korbel and A.M. Rivlin, Micro-analysis of socioeconomic systems: a simulation study, New York 1961, p. 228.
45. Ibid. p. 5.
46. Ibid. p. 7, pp. 42 and 43.

That suggestion can easily be avoided, however, by making use of binary dummy variables. In the above example two of these binary dummy variables are needed. The reason for representing three levels of a qualitative variable in a regression by two binary dummy variables will now be explained in more detail. Suppose that in the regression of a real variable Y on a matrix of real regressors and a constant-term variable one wishes to introduce a qualitative magnitude as a regressor. In this connection one could think, e.g., of a regressor that would have to represent the region. If the data refer to k different regions, the qualitative regressor 'region' could be represented by k different binary dummy variables $\underline{Z}_i$ with $i = 1,2,\ldots,k$. The value of the tth element in $\underline{Z}_i$ is equal to one if Y_t is an observation from the tth region, and is equal to zero in case the tth observation of the variable Y comes from another region. In this design, one and only of the $\underline{Z}_i$'s has the value one for the tth observation.

If matrix $\underline{W}$, consisting of the constant-term variable and the k variables $\underline{Z}_i$, is added to the regressors in the regression of $\underline{Y}$ on $\underline{X}$, one will be confronted with complete multicollinearity. For, summing up the elements of the variables $\underline{Z}_i$ for each of the observations, one obtains the value of the constant-term variable. The complete multicollinearity is eliminated by omitting one of the binary dummy variables, leaving k - 1 binary dummy variables.

Taking up the general argument we may recall that in section 2.2 it has become evident that the regression coefficients from the regression of a real variable $\underline{Y}$ on the real variables from $\underline{X}$ and the dummy variables from $\underline{W}$ can be studied with the help of the estimated disturbances from the regressions of $\underline{Y}$ on $\underline{W}$ and of $\underline{X}$ on $\underline{W}$. As in

section 2.2, the analysis of the regression results for the regression of $\underline{Y}$ on $\underline{X}$ and $\underline{W}$ - this time with a $\underline{W}$-matrix containing the constant-term variable and more than one binary dummy variable - will be preceded by a study of the regression of a real variable on a set of dummy variables.
The linear relation between the real variable $\underline{V}_4$ and the qualitative magnitude, which is expressed by the constant-term variable and the binary dummy variables $\underline{Z}_5$ and $\underline{Z}_6$, can be written as follows:

$$\underline{V}_4 = \underline{1}\gamma_{40} + \underline{Z}_5\gamma_{41} + \underline{Z}_6\gamma_{42} + \underline{\varepsilon}_4 \quad \text{or}$$

$$\begin{bmatrix} V_{41} \\ V_{42} \\ \cdot \\ \cdot \\ \cdot \\ \cdot \\ \cdot \\ \cdot \\ \cdot \\ \cdot \\ \cdot \\ \cdot \\ \cdot \\ \cdot \\ \cdot \\ \cdot \\ \cdot \\ V_{4T} \end{bmatrix} = \begin{bmatrix} 1 & 1 & 0 \\ 1 & 1 & 0 \\ \cdot & \cdot & \cdot \\ \cdot & \cdot & \cdot \\ \cdot & \cdot & \cdot \\ \cdot & 1 & 0 \\ \cdot & 0 & 1 \\ \cdot & 0 & 1 \\ \cdot & \cdot & \cdot \\ \cdot & \cdot & \cdot \\ \cdot & \cdot & \cdot \\ \cdot & \cdot & 1 \\ \cdot & \cdot & 0 \\ \cdot & \cdot & 0 \\ \cdot & \cdot & \cdot \\ \cdot & \cdot & \cdot \\ \cdot & \cdot & \cdot \\ 1 & 0 & 0 \end{bmatrix} \begin{bmatrix} \gamma_{40} \\ \gamma_{41} \\ \gamma_{42} \end{bmatrix} + \begin{bmatrix} \varepsilon_{41} \\ \varepsilon_{42} \\ \cdot \\ \cdot \\ \cdot \\ \cdot \\ \cdot \\ \cdot \\ \cdot \\ \cdot \\ \cdot \\ \cdot \\ \cdot \\ \cdot \\ \cdot \\ \cdot \\ \cdot \\ \varepsilon_{4} \end{bmatrix} \qquad (2.3.1)$$

In (2.3.1) the variables have been arranged in such a way that the first K_1 observations of $\underline{V}_4$ belong to the first category, the next K_2 observations to the second and the remaining $(T-K_1-K_2) = K_3$ observations to the third category. If the relevant means of $\underline{V}_4$ are indicated by $\bar{V}_{41}$, $\bar{V}_{42}$ and $\bar{V}_{43}$, then the estimators for the parameters in (2.3.1) read[47]:

$$\begin{bmatrix} \hat{\gamma}_{40} \\ \hat{\gamma}_{41} \\ \hat{\gamma}_{42} \end{bmatrix} = \begin{bmatrix} \bar{V}_{43} \\ \bar{V}_{41} - \bar{V}_{43} \\ \bar{V}_{42} \quad \bar{V}_{43} \end{bmatrix} \tag{2.3.2}$$

The estimated disturbances belonging to this regression are obtained after substitution of (2.3.2) into (2.3.1) by means of

$$\hat{\varepsilon}_{4t} = V_{4t} - \hat{\gamma}_{40} - \hat{\gamma}_{41} Z_{5t} - \hat{\gamma}_{42} Z_{6t} \tag{2.3.3}$$

The vector of estimated disturbances can be written in extenso as:

$$\hat{\varepsilon}_4 = \begin{bmatrix} V_{41} - \bar{V}_{41} \\ V_{42} - \bar{V}_{41} \\ \vdots \\ V_{4K_1} - \bar{V}_{41} \\ V_{4(K_1+1)} - \bar{V}_{42} \\ V_{4(K_1+2)} - \bar{V}_{42} \\ \vdots \\ V_{4(K_1+K_2)} - \bar{V}_{42} \\ V_{4(K_1+K_2+1)} - \bar{V}_{43} \\ V_{4(K_1+K_2+2)} - \bar{V}_{43} \\ \vdots \\ V_{4T} - \bar{V}_{43} \end{bmatrix} \tag{2.3.4}$$

47. Use has been made of formula (2.1.7) on p. 19, on the understanding that $\underline{Y}$ is replaced with $\underline{V}_4$ and $\underline{X}$ with the matrix of regressors from (2.3.1).

As is clear from (2.3.4), the series of estimated disturbances consists of the original V_{4t}-values expressed in deviation from the corresponding group means. In contrast to the variable $\underline{V}_4$ the estimated disturbances $\hat{\underline{\varepsilon}}_4$ do not show any differences between group means.
The vectors $\underline{Z}_5$ and $\underline{Z}_6$ are not the only way to represent the three categories mentioned above in a regression. The system of dummy variables printed below could also be used by side with the constant-term variable:

$$Z_{7t} = 1 \quad Z_{8t} = 1 \text{ for } t = 1,2,\ldots,K_1$$

$$Z_{7t} = 1 \quad Z_{8t} = 0 \text{ for } t = K_1+1,K_1+2,\ldots,K_1+K_2$$

$$Z_{7t} = 0 \quad Z_{8t} = 0 \text{ for } t = K_1+K_2+1,K_1+K_2+2,\ldots,T$$

Analogously to (2.3.1) the linear relation between $\underline{V}_4$ and the dummy variables reads:

$$\begin{bmatrix} V_{41} \\ V_{42} \\ \cdot \\ \cdot \\ \cdot \\ \cdot \\ \cdot \\ \cdot \\ \cdot \\ \cdot \\ \cdot \\ \cdot \\ \cdot \\ \cdot \\ V_{4T} \end{bmatrix} = \begin{bmatrix} 1 & 1 & 1 \\ 1 & 1 & 1 \\ \cdot & \cdot & \cdot \\ \cdot & \cdot & \cdot \\ \cdot & 1 & 1 \\ \cdot & 1 & 0 \\ \cdot & 1 & 0 \\ \cdot & \cdot & \cdot \\ \cdot & \cdot & \cdot \\ \cdot & 1 & 0 \\ \cdot & 0 & 0 \\ \cdot & 0 & 0 \\ \cdot & \cdot & \cdot \\ \cdot & \cdot & \cdot \\ 1 & 0 & 0 \end{bmatrix} \begin{bmatrix} \gamma'_{40} \\ \gamma'_{41} \\ \gamma'_{42} \end{bmatrix} + \begin{bmatrix} \varepsilon_{41} \\ \varepsilon_{42} \\ \cdot \\ \cdot \\ \cdot \\ \cdot \\ \cdot \\ \cdot \\ \cdot \\ \cdot \\ \cdot \\ \cdot \\ \cdot \\ \cdot \\ \varepsilon_{4T} \end{bmatrix} \qquad (2.3.5)$$

For the least-squares estimator of the parameter vector one gets:

$$\begin{bmatrix} \hat{\gamma}'_{40} \\ \hat{\gamma}'_{41} \\ \hat{\gamma}'_{42} \end{bmatrix} = \begin{bmatrix} \bar{V}_{43} \\ \bar{V}_{42} - \bar{V}_{43} \\ \bar{V}_{41} - \bar{V}_{43} \end{bmatrix} \tag{2.3.6}$$

Contrary to the estimators of the parameters the estimated disturbances of the relation in (2.3.5) do not change by a change in the set of dummy variable used; from (2.3.5) and (2.3.6) this can be verified in a simple way.

The question arises whether this statement remains true for all possible sets of dummy variables showing the same subdivision into categories for the qualitative magnitude concerned. To answer this question, let us start with the most simple system of binary dummy variables indicating a subdivision into three categories[48]:

$$\underline{W}_0 = \begin{bmatrix} 1 & 0 & 0 \\ 1 & 0 & 0 \\ \cdot & \cdot & \cdot \\ \cdot & \cdot & \cdot \\ 1 & 0 & 0 \\ 0 & 1 & 0 \\ 0 & 1 & 0 \\ \cdot & \cdot & \cdot \\ \cdot & \cdot & \cdot \\ 0 & 1 & 0 \\ 0 & 0 & 1 \\ 0 & 0 & 1 \\ \cdot & \cdot & \cdot \\ \cdot & \cdot & \cdot \\ 0 & 0 & 1 \end{bmatrix} \tag{2.3.7}$$

48. This system does not contain a constant-term variable, owing to which the three categories must be represented by means of three instead of two binary dummy variables. In regression analysis it is customary to include the constant-term variable explicity. The systems (2.3.1) and (2.3.5) are, therefore, more suitable than (2.3.7).

For further adstruction, Goldberger[49] introduces a matrix $\underline{P}$ to characterize the transition from one specification of the dummy variables to another. Matrices $\underline{P}_1$ and $\underline{P}_2$, which pre-multiplied by $\underline{W}_0$ lead to the specification of the dummy variables from (2.3.1) and (2.3.5), respectively, look as follows:

$$\underline{P}_1 = \begin{bmatrix} 1 & 1 & 0 \\ 1 & 0 & 1 \\ 1 & 0 & 0 \end{bmatrix} \qquad \underline{P}_2 = \begin{bmatrix} 1 & 1 & 1 \\ 1 & 1 & 0 \\ 1 & 0 & 0 \end{bmatrix} \tag{2.3.8}$$

Regression results which are obtained in a regression on matrix $\underline{W}_0$ from (2.3.7) can serve to produce the estimators of the parameters for any other possible choice of dummy variables, as the regression relating to another specification of the dummy variables, e.g., that from (2.3.1), can be written as:

$$\underline{V}_4 = \underline{W}_0 \underline{P}_1 \underline{\gamma}_4 + \underline{\varepsilon}_4 \tag{2.3.9}$$

In accordance with (2.1.7), see page 19, it follows for $\hat{\underline{\gamma}}_4$:

$$\hat{\underline{\gamma}}_4 = (\underline{P}_1'\underline{W}_0'\underline{W}_0\underline{P}_1)^{-1}\underline{P}_1'\underline{W}_0'\underline{V}_4 \tag{2.3.10}$$

Because $\underline{P}_1$ is an elementary matrix, the inverse exists; the expression of (2.3.10) can be written as

$$\hat{\underline{\gamma}}_4 = \underline{P}_1^{-1}(\underline{W}_0'\underline{W}_0)^{-1}\underline{W}_0'\underline{V}_4 \tag{2.3.11}$$

In the expression $(\underline{W}_0'\underline{W}_0)^{-1}\underline{W}_0'\underline{V}_4$ one recognizes the estimator for the parameter vector from the regression of $\underline{V}_4$ on $\underline{W}_0$; if this is called $\hat{\underline{\gamma}}_0$, the result from (2.3.11) can

---

49. Ibid. p. 7, p. 220.

be simplified into:

$$\underline{\hat{\gamma}}_4 = \underline{P}_1^{-1}\underline{\hat{\gamma}}_0 \tag{2.3.12}$$

It is easy to check that the elements of $\underline{\hat{\gamma}}_0$ have the composition indicated below:

$$\underline{\hat{\gamma}}_0 = \begin{bmatrix} \bar{V}_{41} \\ \bar{V}_{42} \\ \bar{V}_{43} \end{bmatrix} \tag{2.3.13}$$

By calculating the inverse of $\underline{P}_1$ and substituting it into (2.3.12), one get once more the coefficients vector from (2.3.2). Similarly the contents of (2.3.6) can be found back.

Now that the influence of changes in the dummy variables on the corresponding regression coefficients is known, the question about the invariance of the estimated disturbances $\underline{\hat{\varepsilon}}_4$ for the set of dummy variables chosen can be answered.

Starting point is the division into categories as indicated by $\underline{W}_0$. The estimated disturbances resulting from the regression of $\underline{V}_4$ on $\underline{W}_0$ read:

$$\underline{\hat{\varepsilon}}_4 = \underline{V}_4 - \underline{W}_0\underline{\hat{\gamma}}_0 \tag{2.3.14}$$

If matrix $\underline{W}_0\underline{P}_1$ is used instead of $\underline{W}_0$ as a system of dummy variables one gets the estimated disturbances:

$$\underline{\hat{\varepsilon}}_4 = \underline{V}_4 - \underline{W}_0\underline{P}_1\underline{\hat{\gamma}}_4 \tag{2.3.15}$$

By substituting (2.3.12) into (2.3.15) the equality of the expressions of (2.3.14) and (2.3.15) is demonstrated.

Starting from the invariance of $\hat{\underline{\varepsilon}}_4$ for the choice of dummy variables, one can take another look at the regression of a real variable $\underline{Y}$ on the real variables $\underline{X}$ and the dummy variables $\underline{W}_0$. As has been shown in (2.2.19), on page 37, the least-squares estimator of the parameters of the variables from the $\underline{X}$-matrix, to be indicated by $\hat{\underline{\alpha}}_0$, corresponds to the regression coefficients of the regression of the estimated disturbances from the regression of $\underline{Y}$ on $\underline{W}_0$ on the estimated disturbances from the regressions of the variables from $\underline{X}$ on $\underline{W}_0$. As the estimated disturbances from a regression of a variable on $\underline{W}_0$ prove to the invariant for the choice of binary dummy variables, differing from that in $\underline{W}_0$, the same holds for the estimated disturbances $\underline{Y}_C$ and $\underline{X}_C$ from (2.2.19), which implies that $\hat{\underline{\alpha}}_0$ is not dependent of the composition of the dummy variables. On the other hand, the regression coefficients belonging to $\underline{W}_0$ from the regression of $\underline{Y}$ on $\underline{X}$ and $\underline{W}_0$ do depend on the composition chosen for the dummy variables, as may be inferred from (2.2.21) on page 37. If matrix $\underline{W}_0$ is used, the estimators for the elements from the corresponding[50] coefficients vector $\hat{\underline{\delta}}_0$, expressed on the one hand in the group means of the regressand - $\bar{Y}_1$ for the first category, $\bar{Y}_2$ for the second category and $\bar{Y}_3$ for the third category - and on the other hand in the corresponding group means of the variables of $\underline{X}$, $\bar{X}_{1i}$, $\bar{X}_{2i}$ and $\bar{X}_{3i}$ respectively with $i = 1,2,\ldots,\Lambda-1$, read as follows[51]:

50. Formule (2.2.21) is used; because a matrix $\underline{W}_0$ is operated instead of $\underline{W}$, the coefficients $\hat{\underline{\delta}}$ and $\hat{\underline{\alpha}}$ of (2.2.21) are given the index zero.
51. The vector $\hat{\underline{\delta}}_0$ consists of the elements $\hat{\delta}_{0j}$, with $j = 0,1,2$.

$$\hat{\delta}_{00} = \bar{Y}_1 - \sum_{i=1}^{\Lambda-1} \bar{X}_{1i}\hat{\alpha}_{0i}$$

$$\hat{\delta}_{01} = \bar{Y}_2 - \sum_{i=1}^{\Lambda-1} \bar{X}_{2i}\hat{\alpha}_{0i} \qquad (2.3.16)$$

$$\hat{\delta}_{02} = \bar{Y}_3 - \sum_{i=1}^{\Lambda-1} \bar{X}_{3i}\hat{\alpha}_{0i}$$

The coefficients $\hat{\delta}_{00}$, $\hat{\delta}_{01}$ and $\hat{\delta}_{02}$ from (2.3.16) are equal to the constant terms in the regressions of $\underline{Y}$ on $\underline{X}$ for those observations of the variables that belong to, respectively, the first, second and third category, provided that the parameters of $\underline{\alpha}_0$, may be assumed equal for the three categories. One recognizes here the same kind of reservation as in section 2.2 in the comment on formula (2.2.22).

By using, instead of matrix $\underline{W}_0$, the matrix of dummy variables $\underline{W}_0\underline{P}_1$, one obtains, analogously to the derivation explained on pages 51 and 52, as coefficients vector $P_1^{-1}\hat{\underline{Y}}_0$. The elements of $\hat{\underline{\delta}}_1 = \underline{P}_1^{-1}\hat{\underline{\delta}}_0$ are composed as follows[52]:

$$\hat{\delta}_{10} = \bar{Y}_3 - \sum_{i=1}^{\Lambda-1} \bar{X}_{3i}\hat{\alpha}_{0i}$$

$$\hat{\delta}_{11} = (\bar{Y}_1 - \bar{Y}_3) - \sum_{i=1}^{\Lambda-1} (\bar{X}_{1i} - \bar{X}_{3i})\hat{\alpha}_{0i} \qquad (2.3.17)$$

$$\hat{\delta}_{12} = (\bar{Y}_2 - \bar{Y}_3) - \sum_{i=1}^{\Lambda-1} (\bar{X}_{2i} - \bar{X}_{3i})\hat{\alpha}_{0i}$$

---

52. The coefficients of $\hat{\underline{\delta}}_1$ are obtained using (2.3.16), as follows:

$$\begin{bmatrix} \hat{\delta}_{10} \\ \hat{\delta}_{11} \\ \hat{\delta}_{12} \end{bmatrix} = \begin{bmatrix} 1 & 1 & 0 \\ 1 & 0 & 1 \\ 1 & 0 & 0 \end{bmatrix}^{-1} \begin{bmatrix} \hat{\delta}_{00} \\ \hat{\delta}_{01} \\ \hat{\delta}_{02} \end{bmatrix}$$

The last two coefficients $\hat{\delta}_{11}$ and $\hat{\delta}_{12}$ consist in this situation of differences between the constant terms in the abovementioned regressions of $\underline{Y}$ on $\underline{X}$ for a part of the observations.

The numerical outcomes for the estimators of the parameters of the dummy variables in the regression of $\underline{Y}$ on $\underline{X}$ and $\underline{W}$ may show considerable differences among them as a consequence of the freedom of choice concerning the set of binary dummy variables to be used, as will be obvious when (2.3.16) and (2.3.17) are compared. In the analysis of the coefficients of the dummy variables in a regression of a real variable on real variables and dummy variables it is necessary, therefore, to determine the meaning of the coefficients before setting out to interpret the results. A significance test on the parameters, like for instance the one belonging to (2.3.16) and (2.3.17), makes no more sense than it did in section 2.2, if first of all the constancy for all categories of the parameters of the real variables - the vector $\underline{\alpha}_0$ - has not been tested. Chapter 4 will deal with this point in more detail.

For both the relations (2.3.1) and (2.3.5) the significance tests on the parameters correspond to tests worked out beyond the area of regression analysis. The test statistics for the parameters from (2.3.1) look as follows[53]:

---

53. The indices of the test statistics correspond to those of the parameters concerned, from (2.3.1). The test statistics all follow the same t-distribution. The maining of K_1, K_2 and K_3 can be found on page 47.

$$t_{40} = \frac{\bar{V}_{43}}{s\sqrt{\frac{1}{K_3}}}$$

$$t_{41} = \frac{\bar{V}_{41} - \bar{V}_{43}}{s\sqrt{\frac{1}{K_1} + \frac{1}{K_3}}} \qquad (2.3.18)$$

$$t_{42} = \frac{\bar{V}_{42} - \bar{V}_{43}}{s\sqrt{\frac{1}{K_2} + \frac{1}{K_3}}}$$

in which

$$s = \frac{1}{T-3} \sum_{t=1}^{K_1} (V_{4t}-\bar{V}_{41})^2 +$$

$$+ \sum_{t=K_1+1}^{K_1+K_2} (V_{4t}-\bar{V}_{42})^2 + \sum_{t=K_1+K_2+1}^{T} (V_{4t}-\bar{V}_{43})^2 .$$

There is a characteristic difference between the first test statistic and the other two. With the first, the significance of a mean is tested; with the others the significance of differences between means. According to the structure of the matrix containing dummy variables the contents of the test statistics varies. A meaningful interpretation of the above-mentioned significance tests can only be given, therefore, in combination with the meaning of the parameters concerned.

The square of the coefficient of multiple correlation belonging to the regression of a real variable on the binary dummy variables $\underline{W}_0$ or from a matrix obtained from linear combinations of the columns in $\underline{W}_0$, has a simple structure, as Goldberger[54] shows. Applying the formula for R^2 - see (2.1.9) on page 20 - to the regression

54. Ibid. p. 7, pp. 221-222.

calculations for the relation (2.3.1), the estimated disturbances of which are printed under (2.3.4), one gets[55]:

$$R^2=1-\frac{\sum_{t=1}^{K_1}(V_{4t}-\bar{V}_{41})^2+\sum_{t=K_1+1}^{K_1+K_2}(V_{4t}-\bar{V}_{42})^2+\sum_{t=K_1+K_2+1}^{T}(V_{4t}-\bar{V}_{43})^2}{\sum_{t=1}^{T}(V_{4t}-\bar{V}_4)^2}=$$

$$=\frac{K_1(\bar{V}_{41}-\bar{V}_4)^2+K_2(\bar{V}_{42}-\bar{V}_4)^2+K_3(\bar{V}_{43}-\bar{V}_4)^2}{\sum_{t=1}^{T}(V_{4t}-\bar{V}_4)^2} \qquad (2.3.19)$$

From (2.3.19) it appears that R^2 equals zero if all group means are equal. In chapter 3 the relation between the F-test and the R^2 just mentioned is given within the framework of the analysis of variance.

The large number of binary dummy variables introduced by both Lee[56] and Orcutt and others[57] into a regression are a consequence of the wish to use several qualitative magnitudes as explanatory factors for the fluctuations in the regressand. In the context of the regression problem one is examining one might think, for example, that the sex of the individual involved may be an important explanatory factor beside the educational level mentioned on p. 45.

In the two ways to be mentioned below the qualitative magnitude can be included as regressors in the shape of binary dummy variables. After the explanation of both methods the choice between the two will be discussed.

The first method works as follows. The qualitative magnitude 'educational level' (three levels) is

---

55. The mean of variable V_{4t} is denoted by $\bar{V}_4$.
56. Ibid. p. 5.
57. Ibid. p. 45.

represented by means of the constant-term variable and two binary dummy variables; the directions given on p. 46 are thus followed. The qualitative magnitude 'sex' is then represented, according to the same directions, by one binary dummy variable (the last column of $\underline{W}_3$). The matrix with dummy variables will now look as follows:

$$\underline{W}_3 = \begin{bmatrix} 1 & 1 & 0 & 1 \\ 1 & 1 & 0 & 1 \\ . & . & . & . \\ . & . & . & . \\ . & . & . & 1 \\ . & . & . & 0 \\ . & . & . & 0 \\ . & . & . & . \\ . & . & . & . \\ . & 1 & 0 & 0 \\ . & 0 & 1 & 1 \\ . & 0 & 1 & 1 \\ . & . & . & . \\ . & . & . & . \\ . & . & . & 1 \\ . & . & . & 0 \\ . & . & . & 0 \\ . & . & . & . \\ . & . & . & . \\ . & . & 1 & 0 \\ . & . & 0 & 1 \\ . & . & 0 & 1 \\ . & . & . & . \\ . & . & . & . \\ . & . & . & 1 \\ . & . & . & 0 \\ . & . & . & 0 \\ . & . & . & . \\ . & . & . & . \\ 1 & 0 & 0 & 0 \end{bmatrix} \qquad (2.3.20)$$

The second method approaches the structure of the binary dummy variables in another way. The qualitative magnitudes 'educational level' and 'sex' will then give rise to six categories, as the frequency table below will show.

Table 2.1. Frequencies of the characteristics educational level and sex

Educational level \ Sex	Male	Female	Total
Lower	K_{11}	K_{12}	$K_{1.}$
Medium	K_{21}	K_{22}	$K_{2.}$
Higher	K_{31}	K_{32}	$K_{3.}$
Total	$K_{.1}$	$K_{.2}$	K

To represent the six categories from Table 2.1 by dummy variables, one needs, as is shown on p. 46, six dummy variables in all, namely the constant-term variable and five binary dummy variables. The following matrix can be constructed:

$$
\underline{W}_4 = \begin{bmatrix}
1 & 1 & 0 & 1 & 1 & 0 \\
1 & 1 & 0 & 1 & 1 & 0 \\
\cdot & \cdot & \cdot & \cdot & \cdot & \cdot \\
\cdot & \cdot & \cdot & \cdot & \cdot & \cdot \\
\cdot & \cdot & \cdot & 1 & 1 & \cdot \\
\cdot & \cdot & \cdot & 0 & 0 & \cdot \\
\cdot & \cdot & \cdot & 0 & 0 & \cdot \\
\cdot & \cdot & \cdot & \cdot & \cdot & \cdot \\
\cdot & \cdot & \cdot & \cdot & \cdot & \cdot \\
\cdot & 1 & 0 & 0 & \cdot & 0 \\
\cdot & 0 & 1 & 1 & \cdot & 1 \\
\cdot & 0 & 1 & 1 & \cdot & 1 \\
\cdot & \cdot & \cdot & \cdot & \cdot & \cdot \\
\cdot & \cdot & \cdot & \cdot & \cdot & \cdot \\
\cdot & \cdot & \cdot & 1 & \cdot & 1 \\
\cdot & \cdot & \cdot & 0 & \cdot & 0 \\
\cdot & \cdot & \cdot & 0 & \cdot & 0 \\
\cdot & \cdot & \cdot & \cdot & \cdot & \cdot \\
\cdot & \cdot & \cdot & \cdot & \cdot & \cdot \\
\cdot & \cdot & 1 & 0 & \cdot & \cdot \\
\cdot & \cdot & 0 & 1 & \cdot & \cdot \\
\cdot & \cdot & 0 & 1 & \cdot & \cdot \\
\cdot & \cdot & \cdot & \cdot & \cdot & \cdot \\
\cdot & \cdot & \cdot & \cdot & \cdot & \cdot \\
\cdot & \cdot & \cdot & 1 & \cdot & \cdot \\
\cdot & \cdot & \cdot & 0 & \cdot & \cdot \\
\cdot & \cdot & \cdot & 0 & \cdot & \cdot \\
\cdot & \cdot & \cdot & \cdot & \cdot & \cdot \\
\cdot & \cdot & \cdot & \cdot & \cdot & \cdot \\
1 & 0 & 0 & 0 & 0 & 0
\end{bmatrix} \qquad (2.3.21)
$$

The only difference between matrices $\underline{W}_3$ and $\underline{W}_4$ consists in the last two columns of $\underline{W}_4$. The structure of matrix $\underline{W}_3$ implies that the influence of sex on the regressand is supposed to be equal for all educational levels, for the dummy variable of sex is not related to the dummy variables for the educational level, but added to it. On the contrary in $\underline{W}_4$ a separate influence on the regressand is ascribed to each possible combination of the characteristics 'educational level' and 'sex'. The two characteristics are now related to each other, for each number from the last two columns of $\underline{W}_4$ is obtained by multiplying the relevant numbers from, respectively, the

second and fourth column and the third and fourth column. In this connection one speaks of interaction between the two qualitative magnitudes; the last two dummy variables represent the interaction. If in expressing the qualitative magnitudes in terms of dummy variable the former method is used, interaction is supposed to be absent, wrongly perhaps in most cases. The significance of the interaction ought to be tested before a priori such a decision is taken. The second method ought to be followed, therefore. Should the significance test on the interaction[58], expressed by means of the last two columns of matrix $\underline{W}_4$, give no reason to take account of the interaction, then the last two columns of $\underline{W}_4$ can be omitted. In a roundabout way one may thus come back to operating matrix $\underline{W}_3$. The important difference with using the $\underline{W}_3$ matrix in the first place is that now the absence of interaction has been tested rather than assumed.
For translating a number of qualitative quantities into dummy variables the second method should always be preferred. If there are, for example, three qualitative quantities with, respectively, q, r and s categories, then qrs dummy variables are needed, viz. the constant-term variables and qrs-1 binary dummy variables, etc.

2.4. THE BINARY DUMMY VARIABLE AS A REGRESSAND

A binary dummy variable acting as regressand will be used when the magnitude to the 'explained' has a dichotomous character. This is not all a common situation in econometrics. Neter and Maynes[59] in their search through the volumes 1955-1966 of the Journal of the American Statistical Association and the Review of Economics and

58. This test will be discussed in detail in section 3.2.
59. Ibid. p. 6.

Statistics found only nine articles in which a binary dummy variable was acting as a regressand. Among these articles there was one by J.A. Fisher[60], who studied the purchases of durable consumption goods for 762 households in the United States in 1957. Those who bought a durable consumption good in 1957 were subdivided into two categories, viz. those who paid cash and those who bought on credit. For 'explaining' why durable consumption goods were, or were not, bought on credit, the author uses a regressand which gets the value one for cash payment and zero for purchases on credit.
Orcutt and others[61] give an example of a binary regressand with reference to women who do, or do not, work outside their homes. The regressand has the value one for women working outside their homes, and zero for all other cases.
Goldberger[62] makes a few remarks about regressions with qualitative dependent variables. He demonstrates that of the assumptions relevant to the linear model (see section 2.1):

$$\underline{Y} = \underline{X}\underline{\beta} + \underline{\varepsilon} \qquad (2.4.1)$$

the one about homescedastic disturbances cannot be maintained if the regressand has the form of a binary dummy variable. He shows this as follows.
If the tth observation of the regressand amounts to one, then there remains for the disturbance ε_t:

60. J.A. Fisher, An analysis of consumer goods expenditures in 1957, Review of Economics and Statistics, Vol. 44 (1962), pp. 64-71.
61. Ibid. p. 45, pp. 226 ff.
62. Ibid. p. 7, pp. 248 ff.

$$\varepsilon_t = 1 - \sum_{i=1}^{\Lambda} X_{it}\beta_i \qquad (2.4.2)$$

and if Y_t has the value zero:

$$\varepsilon_t = - \sum_{i=1}^{\Lambda} X_{it}\beta_i \qquad (2.4.3)$$

The assumption that the expected value of all elements from $\underline{\varepsilon}$ is equal to zero implies that the distribution of the disturbances for each value of t should look as follows:

$$\begin{aligned} f(\varepsilon_t) &= \sum_{i=1}^{\Lambda} X_{it}\beta_i \qquad \text{for } \varepsilon_t = 1 - \sum_{i=1}^{\Lambda} X_{it}\beta_i \\ f(\varepsilon_t) &= 1 - \sum_{i=1}^{\Lambda} X_{it}\beta_i \quad \text{for } \varepsilon_t = - \sum_{i=1}^{\Lambda} X_{it}\beta_i \end{aligned} \qquad (2.4.4)$$

From the distribution of ε_t it follows for the variance of ε_t:

$$\begin{aligned} E\varepsilon_t^2 &= (1- \sum_{i=1}^{\Lambda} X_{it}\beta_i)^2 (\sum_{i=1}^{\Lambda} X_{it}\beta_i) + \\ &\qquad + (- \sum_{i=1}^{\Lambda} X_{it}\beta_i)^2 (1- \sum_{i=1}^{\Lambda} X_{it}\beta_i) = \\ &= (1- \sum_{i=1}^{\Lambda} X_{it}\beta_i) (\sum_{i=1}^{\Lambda} X_{it}\beta_i) \end{aligned} \qquad (2.4.5)$$

Expression (2.4.5) shows that the variances of the elements of $\underline{\varepsilon}$ depend on the observations in the $\underline{X}$-matrix; the assumption of homoscedastic disturbances may not, therefore, be maintained any longer, so that another estimation method will have to be used to estimate $\underline{\beta}$. The

estimation method of generalized least squares[63], which allows of heteroscedasticity in the disturbances, cannot be applied directly, because the covariance matrix of the disturbances contains the unknown parameter vector $\underline{\beta}$. An iterative generalized least-squares method[64] will finally lead to a theoretically justified estimation of $\underline{\beta}$.
If, in the case of a binary dummy variable as regressand, $\underline{\beta}$ would be estimated with the method of least-squares, these coefficients are biased. Neter and Maynes[65] have examined the properties of the correlation coefficient in the regression of a binary dummy variable on a real variable. They point out in their article that the correlation coefficient indicates the degree of correspondence between the regression line and the scatter of points, and that in the case of a binary regressand the scatter consists of two lines of points, viz. one on the level zero and one on the level one; figure 2 (see next page) represents as an illustration the scatter diagram of a binary dummy regressand and a real variable as a regressor. It will be noticed that there is not much sense in adjusting a striaght line to the scatter of points, a fact that also reduces the importance of the degree of adjustment expressed by the correlation coefficient.
The authors recommend, should the case arise, determining the degree of connection not by means of the correlation coefficient but by means of the correlation ratio[66]: the latter criterion does not impose any linearity on the character of the connection between the variables involved.

---

63. See p. 18.
64. See Goldberger, ibid. p. 7, pp. 244 and 245.
65. Ibid. p. 6.
66. See for the meaning of this ratio Yule and Kendall, ibid. p. 2, p. 256.

Figure 2. Scatter diagram of the binary dummy variable Y_t as regressand and the real variable X_t as regressor.

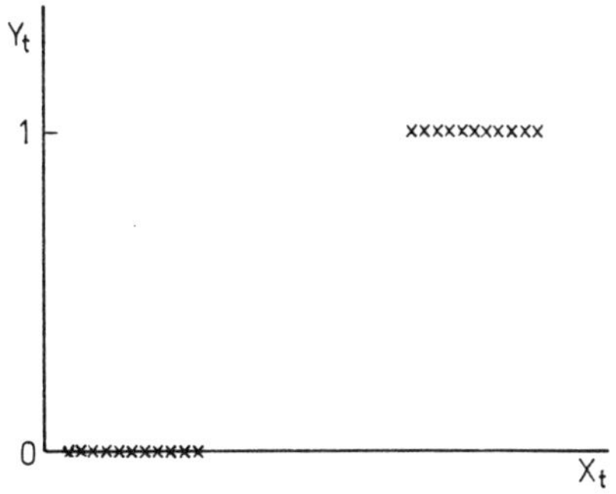

When an estimate for $\underline{\beta}$ has been obtained and if the degree of connection between regressand and regressors is considered sufficient, a more detailed study of the meaning of the estimated regressand values follows. The expectation of the tth element of $\underline{Y}$ reads, in accordance with the relation of (2.4.1):

$$EY_t = \sum_{i=1}^{\Lambda} x_{it}\beta_i \qquad (2.4.6)$$

The meaning of the expected value of Y_t can be derived from (2.4.4). For given values of $X_{1t}, X_{2t}, \ldots, X_{\Lambda t}$, the first line of (2.4.4) gives the probability of the disturbance having the value ε_t at a value of one of Y_t. In the second line of (2.4.4) gives the probability of the value ε_t at a zero value of Y_t. As the probability of obtaining ε_t appears to be related to the value of Y_t, the expression $\sum_{i=1}^{\Lambda} X_{it}\beta_i$ is also equal to the probability of Y_t being 1; similarly $1 - \sum_{i=1}^{\Lambda} X_{it}\beta_i$ is equal to the probability that $Y_t = 0$. Accordingly, the estimated regression values are generally interpreted as the probabilities that $Y_t = 1$.

A serious objection against interpretation of estimated $\underline{Y}$-values as probabilities may be raised because of the possibility that the value of $\hat{Y}_t$ falls outside the closed interval 0-1. As an illustration an example given by Goldberger[67] may serve. He points out that De Zanosi has attempted to explain for the United States of North America, using a sample of individuals, the purchase or non-purchase of a new car in 1952 from the disposable income X_t (measured in thousands of dollars). He carried out a regression with a binary dummy variable Y_t acting as regressand; the variable was supposed to be one when a new car had been bought, and zero when no purchase had been made.

The estimated relation was:

$$\hat{Y}_t = - .008 + .022 \; X_t \qquad (2.4.7)$$

Applying relation (2.4.7) to individuals with a disposable income of more than \$ 45,000 one obtains an estimated regression value that is larger than one. Negative or very low disposable incomes yield negative values. These cases evidently give results that cannot be indicated as probabilities.

The difficulty of interpreting $\hat{Y}_t$ as a probability does not occur in a number of other estimation methods. The relation between the linear relation to be estimated $\sum_{i=1}^{\Lambda} X_{it}\beta_i$ and the probability that $Y_t = 1$ can be established in several ways. Johnson[68] gives a survey of a number of methods that explicity take the restricted variation area of the regressand into account. Because no

67. Ibid. p. 7, p. 249.
68. T. Johnson, Qualitative and limited dependent variables in economic relationships, Econometrica, Vol. 40 (1972), No. 3, pp. 455-462.

use is made of dummy variables, this survey is not further dealt with in this volume.

The weak points (see p. 63) of the least-squares estimator for $\underline{\beta}$ from (2.4.1), added to the abovementioned objection to the common interpretation of $\hat{Y}_t$ as a probability, lead to the conclusion that the estimation method of least-squares is unsuitable to be applied to a linear relation with a binary regressand.

Nevertheless Tintner[69] presents an estimation of parameters that technically corresponds with the execution of a regression on a linear relation with a binary regressand; he does so in the setting of the so-called discriminant analysis. Before going into this point, some explanation of discriminant analysis seems in order.

Discriminant analysis can be applied in the following case[70]. For a sample of consumers who had bought an article on the instalment system, the data of the real variables income, price of the article bought, down payment and credit instalment were collected for each buyer. Furthermore the identity was known of those who had paid the instalments according to the schedule agreed upon, and of those who had failed in that respect. On the ground of these data it was the intention to determine a linear relation, the so-called discriminant function, which was supposed to separate as clearly as possible the category of good payers from that of bad payers.

It will be demonstrated that the parameters in the discriminant function cannot be estimated by means of a regression on a linear relation with a binary regressand without implicitly interfering with the reason for using discriminant analysis at all.

In discriminant analysis the two categories are separated

69. Ibid. p. 6.
70. See G. Tintner, ibid. p. 6, p. 98.

as follows: if the value of the discriminant function, in the antecedents income, price of the article, down payment and credit instalment, exceeds a certain constant c, the individual involved is allotted to the category of bad payers; if not, to the category of good payers.
Once the coefficients from the discriminant function are available, an attempt can be made to predict for a new client presenting himself, on the basis of his antecedents, whether he will be a good or a bad debtor.
The theoretical derivation of the discriminant function, which allots as correctly as possible the individuals of the sample to either category, is found in Kendall and Stuart[71]. They proceed as follows for n real variables. They assume that the tth observation ($t = 1,2,\ldots,T$) of the n real variables, to be indicated by the $n \times 1$ vector $\underline{S}_t$[72], is a random drawing from an n-dimensional normal distributed variable $\underline{S}$. For observations regarding individuals from the first category this normal distribution is characterized by the vector of expected values $\underline{\mu}_1$ and the covariance matrix $\underline{\Omega}$[73]. For the second

71. Ibid. p. 8, pp. 315-316.
72. This vector contains the values of the n variables for the tth observation.

$$\underline{S}_t = \begin{bmatrix} S_{1t} \\ S_{2t} \\ \vdots \\ S_{nt} \end{bmatrix}$$

73.

$$E\underline{S}_t = \begin{bmatrix} ES_{1t} \\ ES_{2t} \\ \vdots \\ ES_{nt} \end{bmatrix} = \begin{bmatrix} \mu_{11} \\ \mu_{21} \\ \vdots \\ \mu_{n1} \end{bmatrix} = \underline{\mu}_1, \quad E(\underline{S}_t-\underline{\mu}_1)(\underline{S}_t-\underline{\mu}_1)' = \underline{\Omega}$$

category the parameters are $\underline{\mu}_2$ and the same $\underline{\Omega}$.
The authors go on to show that if the error made in allotting an individual from the first category to the second is given the same weight as a similar error in the inverse direction, and if moreover one wishes to minimize the total number of such errors, the discriminant function must be derived from the relation between densities described by:

$$\frac{\dfrac{1}{(2\pi)^{n/2}|\underline{\Omega}|^{1/2}} \exp\left[- 1/2\ (\underline{S}-\underline{\mu}_1)'\underline{\Omega}^{-1}(\underline{S}-\underline{\mu}_1)\right]}{\dfrac{1}{(2\pi)^{n/2}|\underline{\Omega}|^{1/2}} \exp\left[- 1/2\ (\underline{S}-\underline{\mu}_2)'\underline{\Omega}^{-1}(\underline{S}-\underline{\mu}_2)\right]} = c \quad (2.4.8)$$

(c is a constant which follows from the distribution and weighting of the two kinds of errors).
The relation of (2.4.8) can be written as:

$$\underline{S}'\underline{\Omega}^{-1}(\underline{\mu}_1-\underline{\mu}_2) = c' \quad (2.4.9)$$

(c' is a constant in which other elements beside c from (2.4.8) play a role).
The left term of (2.4.9) is the discriminant function: the parameters in this function are $\underline{\Omega}^{-1}(\underline{\mu}_1-\underline{\mu}_2)$.
The remaining part of the present section will be devoted to the obtaining of estimators for these parameters.
Kendall and Stuart[74] estimate the parameters in the discriminant function as described below for a sample of T individuals, T_1 of whom belong to the first category and T_2 to the second category:

74. Ibid. p. 8, p. 316.

$$\underline{\hat{\Omega}}^{-1}(\underline{\hat{\mu}}_2-\underline{\hat{\mu}}_1) = \begin{bmatrix} \hat{\omega}_{11} & \hat{\omega}_{12} \cdots & \hat{\omega}_{1n} \\ \hat{\omega}_{12} & \hat{\omega}_{22} \cdots & \hat{\omega}_{2n} \\ \vdots & \vdots & \vdots \\ \hat{\omega}_{1n} & \hat{\omega}_{2n} \cdots & \hat{\omega}_{nn} \end{bmatrix}^{-1} \begin{bmatrix} \bar{S}_{11}-\bar{S}_{12} \\ \bar{S}_{21}-\bar{S}_{22} \\ \vdots \\ \bar{S}_{n1}-\bar{S}_{n2} \end{bmatrix} \qquad (2.4.10)$$

where

$\bar{S}_{i1}$: the mean of the ith variable ($i = 1,2,\ldots,n$) calculated over the T_1 persons of the first category

$\bar{S}_{i2}$: the mean of the ith variable ($i = 1,2,\ldots,n$) calculated over the T_2 persons of the second category

$\hat{\omega}_{ij}$: the so-called pooled covariance of the variables ($i, j = 1,2,\ldots,n$), which for large values of T may be written as

$$\left[\frac{1}{T} \sum_{t=1}^{T_1}(S_{it}-\bar{S}_{i1})(S_{jt}-\bar{S}_{j1})+\sum_{t=T_1+1}^{T}(S_{it}-\bar{S}_{i2})(S_{jt}-\bar{S}_{j2})\right]$$

When calculating the estimator of the covariance matrix the authors explicitly take into account, and rightly so, the differences between the means of the different categories, as can be seen from the formula for $\hat{\omega}_{ij}$. Tintner[75] arrives at estimators for the parameters in the discriminant function that correspond with the regression coefficients $\hat{\theta}_1,\hat{\theta}_2,\ldots,\hat{\theta}_n$ from the following linear relation with a binary regressand:

$$Y_t = (S_{1t}-\bar{S}_1)\theta_1+(S_{2t}-\bar{S}_2)\theta_2+\ldots+(S_{nt}-\bar{S}_n)\theta_n + \varepsilon_t \qquad (2.4.11)$$

---

75. Ibid. p. 6, pp. 97 and 100.

where in this case:

Y_t: binary regressand with a value of $1/T_1$ for the T_1 observations of the first category and $-1/T_2$ for the T_2 observations of the second category

$\bar{S}_i$: the mean of the ith variable $(i = 1,2,\ldots,n)$ calculated over all T observations

ε_t: disturbance term.

The least-squares estimators for θ_i read (formula (2.1.7) on page 19, being applied):

$$\begin{bmatrix} \hat{\theta}_1 \\ \hat{\theta}_2 \\ \vdots \\ \hat{\theta}_n \end{bmatrix} = \begin{bmatrix} \hat{\omega}'_{11} & \hat{\omega}'_{12} \ldots \hat{\omega}'_{1n} \\ \hat{\omega}'_{12} & \hat{\omega}'_{22} \ldots \hat{\omega}'_{2n} \\ \vdots & \vdots \quad\quad \vdots \\ \hat{\omega}'_{1n} & \hat{\omega}'_{2n} \ldots \hat{\omega}'_{nn} \end{bmatrix}^{-1} \begin{bmatrix} \bar{S}_{11} - \bar{S}_{12} \\ \bar{S}_{21} - \bar{S}_{22} \\ \vdots \\ \bar{S}_{n1} - \bar{S}_{n2} \end{bmatrix} \qquad (2.4.12)$$

in which

$$\hat{\omega}'_{ij} = \frac{1}{T} \sum_{t=1}^{T} (S_{it}-\bar{S}_i)(S_{jt}-\bar{S}_j)$$

When the estimators for θ_i from (2.4.12) are compared with the estimated parameters from (2.4.10), the only difference appears that between $\hat{\omega}_{ij}$ and $\hat{\omega}'_{ij}$. In calculating $\hat{\omega}_{ij}$ a distinction is made between the categories by operating different mean values. Tintner's approach looks at first sight inconsistent in applying the regression coefficients from (2.4.12) in that the makes use of an assumed equality $\bar{S}_{i1} = \bar{S}_{i2} = \bar{S}_i$ for all i when calculating the estimators for the elements of $\underline{\Omega}$. Should this assumption be true of the population, then there would be no reason for discriminant analysis; there would be no distinct categories since group means as well as dispersions would be the same. However using the regression (2.4.12) to estimate the coefficients of the discriminant function is adequate if the constant c' of (2.4.9) is adjusted properly[76].

2.5. SOME REMARKS ABOUT THE USE OF BINARY DUMMY VARIABLES IN TIME-SERIES ANALYSIS

In the econometric quarterly model for the United Kingdom Klein and other[76] made an extensive use of binary dummy variables to correct for seasonal fluctuations; they called this kind of variable quarterly variables. The authors of the model pay rather a lot of attention to the treatment of seasonal movements in time series, achieving a method which has been copied in several other models[77] since. Before discussing the method of the quarterly variables we will recapitulate the objections that first of all Klein and others[78], but also R.J. Wonnacott and T.H. Wonnacott[79] and Thomas and Wallis[80] have raised against the elementary methods of eliminating seasonal effects from time series[81].

---

76. Ibid. p. 7, p. 42.
77. See, e.g., D. Lüdeke, Ibid. p. 7; the Japanese model by L.R. Klein and Y. Shinkai, Ibid. p. 6; the quarterly model for Japan by S. Ichimura, L.R. Klein, S. Koizuma, K. Sato and Y. Shinkai, A quarterly econometric model of Japan, 1952-1959, Osaka Economic Papers, Vol. 12 (1962). In the introduction to the Brookings model, ibid. p. 6, p. 30, a preference for quarterly variables is equally expressed.
78. Ibid. p. 7, pp. 40-44.
79. R.J. Wonnacott and T.H. Wonnacott, Introductory Statistics, New York 1969, pp. 277-278.
80. J.J. Thomas and K.F. Wallis, Seasonal variation in regression analysis, Journal of the Royal Statistical Society (Series A), Vol. 134 (1971), pp. 57-73.
81. See, e.g., R.C. Clelland, J.S. de Cani, F.E. Brown, J.P. Bursk and D.S. Murray, Basic statistics with business applications, New York 1966, Chapter 13.

(i) Estimating the seasonal patterns is done in most cases without using a probability distribution (the so-called non-parametric methods). This implies that one does not know the degrees of freedom used up in the estimation procedure and so cannot correct for them in later calculations. With more complex smoothing techniques it is even likely to have lost track of what one has corrected for after having eliminated the so-called seasonal fluctuations.

(ii) If one of the techniques of moving averages is used to obtain a series that is corrected for seasonal fluctuations, that series will show autoregression, even if the uncorrected series did not.

(iii) Unless the same technique is applied throughout for the correction of seasonal fluctuations - which is hardly likely when series composed by different collecting agencies are used - there is no insight whatsoever in the influence of the corrections on the mutual relations between the series.

(iv) Elementary methods of eliminating seasonal influences in many cases imply a step-by-step determination of the seasonal pattern, for the trend movement is assessed first, and the seasonal pattern next. In a time series, however, both movements manifest themselves simultaneously, which seems to present an argument for simultaneous establishment of both components.

In view of the disadvantages attached to (elementary) methods of eliminating seasonal influences, Klein et al., among others, advise against the use of time series pre-corrected for seasonal movements for estimating the

parameters in a model. Klein et al., have developed the procedure to be described below, which meets the first three objections and for which, as will become evident later on, the fourth objection is not relevant.
Into each equation of the quarterly model the binary dummy variables Q_{it} are introduced as regressors, i being 1, 2, 3, and

Q_{it} = 1 for observations of the ith quarter
Q_{it} = 0 for observations outside the ith quarter.

A linear relation between the variables Y_t and X_{1t} will read after the addition of the quarterly variables:

$$Y_t = \delta_0 + \delta_1 Q_{1t} + \delta_2 Q_{2t} + \delta_3 Q_{3t} + \alpha_1 X_{1t} + \varepsilon_{1t} \qquad (2.5.1)$$

The relation (2.5.1) is based on the assumption of additivity and linearity of the seasonal influences. The composers of the English quarterly model furthermore give the following comment on (2.5.1)[82]:

'It implies that the relation between Y_t and X_{1t} makes parallel shifts in position from quarter to quarter. The constant term[83] in the equation can assume four possible values. Further degrees of complication to bring out other forms of seasonal influence can be introduced if necessary. For example, if the slope, as well as the level, of the equation were to shift with the seasons, we could write

---

82. Ibid. p. 7, p. 43, The symbols in the quatation are changed according to those of (2.5.1).
83. It is meant to say $\delta_0 + \sum_{i=1}^{3} \delta_i Q_{it}$.

$$Y_t = \delta_0 + \delta_1 Q_{1t} + \delta_2 Q_{2t} + \delta_3 Q_{3t} + \alpha_1 X_{1t} + \alpha_2 Q_{1t} X_{1t} + \\ + \alpha_3 Q_{2t} X_{1t} + \alpha_4 Q_{3t} X_{1t} + \varepsilon_{2t} \qquad (2.5.2)$$

The interaction variables $Q_{it}X_{1t}$ assume non-zero values in i^{th} quarters of each year and change[84] the slope coefficient of X_{1t} by the amount $\alpha_{(i+1)}$. In our empirical work we have followed a single pattern throughout and assumed that only the constant term in each equation is subject to seasonal variation. This is the simplest approach, but it was selected only after examination of a large number of graphs of relationships in the model. In these graphs, the seasonal scatter of points conforms clearly to a pattern of parallel layers of relationships. The slope coefficients appear to have no significant seasonal variation.'

The parameters δ_1, δ_2 and δ_3, as well as α_2, α_3 and α_4, express the seasonal movement, for they represent the shifts in the parameters δ_0 and α_1 of the fourth quarter. The estimation of these parameters in (2.5.1) requires three degrees of freedom, that of the one in (2.5.2) requires six. The first objection against elementary methods of elimination of seasonal fluctuations has thus been remedied. The second objection is met by not using moving averages in determining seasonal influences. The third objection is disposed of because for X and Y the same correction method is used. The fourth objection, directed against the step-by-step determination of the components of a time series, does not apply to the method of Klein et al. because they do not make a subdivision

84. By 'change' is meant: change with respect to the slope coefficient of the fourth quarter (α_1).

into trend movement and seasonal influence. This can be explained as follows. The coefficients of the dummy variables in (2.5.1) can be expressed (compare (2.3.17) p. 54) in the quarterly means of the real variables Y_t and X_{1t}, to be denoted by $\bar{Y}_j$ and $\bar{X}_{1j}$ ($j = 1,2,3,4$), thus:

$$\begin{bmatrix} \hat{\delta}_0 \\ \hat{\delta}_1 \\ \hat{\delta}_2 \\ \hat{\delta}_3 \end{bmatrix} = \begin{bmatrix} \bar{Y}_4 - \hat{\alpha}_1 \bar{X}_{14} \\ \bar{Y}_1 - \bar{Y}_4 - \hat{\alpha}_1 (\bar{X}_{11} - \bar{X}_{14}) \\ \bar{Y}_2 - \bar{Y}_4 - \hat{\alpha}_1 (\bar{X}_{12} - \bar{X}_{14}) \\ \bar{Y}_3 - \bar{Y}_4 - \hat{\alpha}_1 (\bar{X}_{13} - \bar{X}_{14}) \end{bmatrix} \qquad (2.5.3)$$

If the time series of Y_t and X_{1t} are conceived of as built up of the components 'trend' and 'season', with 'accidental' and 'once-only' movements added, and if the two last-mentioned components are moreover supposed to be, on an average, negligible for each quarter, then the mean values $\bar{Y}_j$ and $\bar{X}_{1j}$ will contain both the trend value and the seasonal component. Apart from the fact that the coefficients of the Q_{it}-variables contain more than the seasonal component there are other objections to be made against the author's method as described in the quotation on p. 75. First, additivity and linearity of the seasonal influences are assumed; if the pattern is otherwise, the method described is unsuitable. Only for a multiplicative pattern can the procedure be applied after a logarithmic transformation. Second, the coefficients for the quarterly variables can represent shifts in the constant only if in (2.5.1) α_1 may be considered constant over the four quarters. Third, the test for constancy of α_1 as visualized by Klein et al., can function only if no more than one real regressor figures in the relation. With more than one real variable as regressors, simultaneous testing for constancy is required.

Finally, Klein's method is subjective, for the degree of parallelism of lines would have to be determined at sight. The method to be developed in chapter 4 to test the constancy of parameters meets these objections, except the first one, to Klein's method.

2.6. USING BINARY DUMMY VARIABLES IN SITUATIONS OF ERRORS IN VARIABLES

As an introduction the most simple situation of errors in variables is to be described, after which the common corresponding estimating methods will be mentioned.[85] Next it will be demonstrated that Wald's estimator[86] can be obtained with the help of a binary dummy variable operated as an instrumental variable. Afterwards the approach by instrumental variables will be generalized to cases with more than one instrumental variable, or more than one regressor. Bartlett's estimator[87] is presented as a special case of the application of two binary dummy variables as instrumental variables for one regressor. At the end of this section the usefulness of applying dummy variables as instrumental variables will be examined critically.

The most simple situation can be represented as follows:

$$Y_t = \alpha + \beta \tilde{X}_t + u_t \qquad t = 1,2,\ldots,T \tag{2.6.1}$$

$$X_t = \tilde{X}_t + v_t \tag{2.6.2}$$

---

85. Use is made of J. Johnston, ibid. p. 8, chapter 9.
86. A. Wald, The fitting of straight lines if both variables are subject to error, annals of mathematical statistics, Vol. 11 (1940), pp. 284-300.
87. M.S. Bartlett, Fitting a straight line when both variables are subject to error, Biometrics, Vol. 5 (1949), pp. 207-212.

where $\tilde{X}_t$ indicates the true value and X_t the observed value. Further it is assumed as to the stochastic character:

$$Eu_t = Ev_t = 0, \; Eu_t^2 = \sigma_u^2, \; Ev_t^2 = \sigma_v^2 \text{ and } Eu_t v_t = 0.$$

Errors in the variable Y have been relegated to u, $\tilde{X}$ is considered non-stochastic.
Relation (2.6.1) reads in terms of observable quantities:

$$Y_t = \alpha + \beta X_t + (u_t - \beta v_t) \qquad (2.6.3)$$

The estimating method of least-squares results in an asymptotic bias of:

$$\underset{T\to\infty}{\text{plim}}\; \hat{\beta} - \beta = - \frac{\sigma_v^2 \beta}{\sigma_{\tilde{X}}^2 + \sigma_v^2} \qquad (2.6.4)$$

where $\sigma_{\tilde{X}}^2 = \frac{1}{T} \sum_{t=1}^{T} (\tilde{X}_t - \bar{\tilde{X}})^2$.

With the intention to eliminate this asymptotic bias three approaches have been followed, viz.:

a. the method of maximum likelihood;
b. the method of instrumental variables;
c. the grouping method.

For the first-mentioned approach the distributions of u and v as well as the relation between the corresponding variances must be known a priori. Because obtaining this kind of information is rather doubtful in practice - notably for cases with more than one regressor for which the a priori information needed increases rapidly - this approach will not be discussed any further here.

The instrumental variable method will now be described in more detail for the above simple case, after which it will be deomstrated that Wald's grouping method can be conceived of as a special case of the instrumental variable method. For relation (2.6.1), to achieve a consistent estimator for β, an instrumental variable W must be available, such that the following conditions are fulfilled:

1. $\underset{T\to\infty}{\text{plim}} \frac{1}{T} \sum_{t=1}^{T} (W_t-\bar{W})(u_t-\beta v_t) = 0$
2. $\underset{T\to\infty}{\text{plim}} \frac{1}{T} \sum_{t=1}^{T} (W_t-\bar{W})^2 = \sigma_W^2 \qquad 0 < \sigma_W^2 < \infty$
3. $\underset{T\to\infty}{\text{plim}} \frac{1}{T} \sum_{t=1}^{T} (W_t-\bar{W})(X_t-\bar{X}) = \sigma_{WX} \qquad |\sigma_{WX}| < \infty \text{ and } \sigma_{WX} \neq 0$

Once such a variable is available, it is true of the estimator

$$b = \frac{\sum_{t=1}^{T} (W_t-\bar{W})(Y_t-\bar{Y})}{\sum_{t=1}^{T} (W_t-\bar{W})(X_t-\bar{X})} \qquad (2.6.5)$$

that $\underset{T\to\infty}{\text{plim}}\, b = \beta$ and the asymptotic variance of $b = \frac{1}{T}\sigma^2_{(u-\beta v)} \frac{\sigma_W^2}{\sigma_{WX}^2}$. The asymptotic variance of b is estimated by means of

$$s^2 \frac{\sum_{t=1}^{T} (W_t-\bar{W})^2}{\left[\sum_{t=1}^{T} (W_t-\bar{W})(X_t-\bar{X})\right]^2},$$

in which

$$s^2 = \frac{1}{T-2} \sum_{t=1}^{T} (Y_t-a-bX_t)^2. \qquad (2.6.6)$$

Now it will be shown that Wald's estimator can be conceived of as an application of the method of instrumental variables just described. Wald groups the T observations X_t according to size. The mean of the first group of T/2 smallest observations of the X variable is denoted by $\bar{X}_1$; for the corresponding observations of the variable Y_t the symbol $\bar{Y}_1$ is used. For the second group the means are indicated by $\bar{X}_2$ and $\bar{Y}_2$, respectively. Wald's estimator reads:

$$b_W = \frac{\bar{Y}_2 - \bar{Y}_1}{\bar{X}_2 - \bar{X}_1} \qquad (2.6.7)$$

It can easily be checked that (2.6.7) results from (2.6.5) if the instrumental variable W assumes the value 0 for observations of X from the first group and the value 1 for those from the second group. In a later stage the question whether such a choice of W fulfils conditions 1 and 3 on p. 79 will be taken up again; condition 2 does not present any difficulties in this context.

For the generalization[88] of the method of instrumental variables to include cases with more than one regressor we shall examine the linear relation (2.1.2) on page 17, after expressing the variables in terms of deviation from their mean and after translation into matrix notation.

$$\underline{Y} = \underline{\underline{X}}\underline{\beta} + \underline{\varepsilon} \qquad (2.6.8)$$

where:

- $\underline{Y}$: vector with T observations of the regressand
- $\underline{X}$: matrix with T observations of each of the $(\Lambda-1)$ regressors
- $\underline{\beta}$: vector with $(\Lambda-1)$ parameters

---

88. What follows can in part be found back in J. Johnston, ibid. p. 8, pp. 279 ff.

$\underline{\varepsilon}$: vector with T disturbacnce terms, $E\underline{\varepsilon} = \underline{0}$ and $E\underline{\varepsilon}\underline{\varepsilon}' = \sigma_\varepsilon^2 \underline{I}$.

Because of observational errors it is no longer true that $\underline{X}$ and $\underline{\varepsilon}$ are uncorrelated and therefore the OLS estimators for $\underline{\beta}$ lack consistency. In this situation one could achieve consistent estimators for $\underline{\beta}$ if one had a minimum of $(\Lambda-1)$ instrumental variables at one's disposal such that it would be true that (with $\underline{W}$ the matrix of instrumental variables):

1. $\underset{T\to\infty}{\text{plim}} \frac{1}{T} \underline{W}'\underline{\varepsilon} = \underline{0}$
2. $\underset{T\to\infty}{\text{plim}} \frac{1}{T} \underline{W}'\underline{W}$ exists and its inverse can be taken
3. $\underset{T\to\infty}{\text{plim}} \frac{1}{T} \underline{W}'\underline{X}$ exists and has full rank

When the instrumental variable method is used, relation (2.6.8) is observed after pre-multiplication by $\underline{W}'$

$$\underline{W}'\underline{Y} = \underline{W}'\underline{X}\underline{\beta} + \underline{W}'\underline{\varepsilon} \tag{2.6.9}$$

If $\underline{W}$ is of the order $T\times(\Lambda-1)$ and contains, therefore, as many instrumental variables as $\underline{X}$ contains regressors, then the estimator for $\underline{\beta}$ reads:

$$\underline{\hat{\beta}} = (\underline{W}'\underline{X})^{-1}\underline{W}'\underline{Y} \tag{2.6.10}$$

The corresponding asymptotic covariance matrix reads:

$$\sigma_\varepsilon^2 \underset{T\to\infty}{\text{plim}} \frac{1}{T} \left(\frac{1}{T} \underline{W}'\underline{X}\right)^{-1} \left(\frac{1}{T} \underline{W}'\underline{W}\right) \left(\frac{1}{T} \underline{X}'\underline{W}\right)^{-1} \tag{2.6.11}$$

If $\underline{W}$ contains more than $(\Lambda-1)$ instrumental variables, the common least-squares method can be applied to the system of equations (2.6.9), resulting into the following estimator for $\underline{\beta}$[89]:

$$\hat{\underline{\beta}} = (\underline{X}'\underline{W}\underline{W}'\underline{X})^{-1}\underline{X}'\underline{W}\underline{W}'\underline{Y} \qquad (2.6.12)$$

with the asymptotic covariance matrix

$$\sigma_{\varepsilon}^{2} \operatorname*{plim}_{T\to\infty} \frac{1}{T}(\frac{1}{T}\underline{X}'\underline{W}\frac{1}{T}\underline{W}'\underline{X})^{-1}\frac{1}{T}\underline{X}'\underline{W}\frac{1}{T}\underline{W}'\underline{W}\frac{1}{T}\underline{W}'\underline{X}(\frac{1}{T}\underline{X}'\underline{W}\frac{1}{T}\underline{W}'\underline{X})^{-1} \qquad (2.6.13)$$

As will be demonstrated presently, the Bartlett estimator can be conceived of as a special case of the estimator (2.6.12). The disturbance vector in (2.6.9) has no longer a covariance matrix of the shape $\sigma^2\underline{I}$, so that it would be better to apply the method of generalized least-squares to (2.6.9). The result would be as follows:

$$\hat{\underline{\beta}} = [\underline{X}'\underline{W}(\underline{W}'\underline{W})^{-1}\underline{W}'\underline{X}]^{-1}\underline{X}'\underline{W}(\underline{W}'\underline{W})^{-1}\underline{W}'\underline{Y} \qquad (2.6.14)$$

and the relevant covariance matrix reads:

$$\operatorname*{plim}_{T\to\infty} \frac{1}{T}\,[\frac{1}{T}\,\underline{X}'\underline{W}(\frac{1}{T}\underline{W}'\underline{W})^{-1}\frac{1}{T}\,\underline{W}'\underline{X}]^{-1} \qquad (2.6.15)$$

Estimator (2.6.14) can also be looked upon as the result of a two-stage procedure. The first step consists in the regression of $\underline{X}$ on $\underline{W}$ resulting in the determination of $\hat{\underline{X}}$. The second step consists in the regression of $\underline{Y}$ on $\hat{\underline{X}}$. Elsewhere has been discussed in more detail the relation between the use of dummy variables as instrumental variables and simultaneous estimating methods applied to a model composed of a relation like (2.6.3) and an

89. It should be borne in mind that the variables $\underline{W}'\underline{Y}$ and $\underline{W}'\underline{X}$ are to be expressed in deviation from their mean.

auxiliary equation with dummy regressors[90].

It will now be demonstrated that Bartlett's estimator is an application of estimator (2.6.12) with $\underline{W}$ consisting of binary dummy variables[91]. Again the starting point will be relation (2.6.1) with the relevant assumptions. In observable quantities this relation reads - see (2.6.3) - :

$$Y_t = \alpha + \beta X_t + (u_t - \beta v_t) \tag{2.6.16}$$

Bartlett subdivides the observations X_t into three groups of equal numbers of elements. After the X_t's have been arranged according to size, the first T/3 observations are allotted to the first group, the second series to the second group and the remainder to the third. The corresponding group means are indicated by $\bar{X}_1$, $\bar{X}_2$ and $\bar{X}_3$. The observations Y_t are also divided into three groups in accordance with the subdivision of the corresponding X_t's, the group means to be denoted by $\bar{Y}_1$, $\bar{Y}_2$ and $\bar{Y}_3$. Bartlett uses as estimator for β:

$$b_B = \frac{\bar{Y}_3 - \bar{Y}_1}{\bar{X}_3 - \bar{X}_1} \tag{2.6.17}$$

The approach with dummy variables as instrumental variables may develop as follows. The division into three groups can be represented with dummy variables using the constant-term variable and two binary dummy variables,

---

90. See M.A. Kooyman, The use of dummy variables in linear relations, paper presented at the European Econometric Society Meeting 1972, pp. 3-7.
91. J. Johnston, ibid. p. 8, describes in footnote 1 on page 285 that also a three-valued dummy variable with value 1 for observations in the first group, 0 for those in the second group and -1 for those in the third group can be used. His remark can be verified in a simple way with the help of the explanation on p. 79 of this volume. Here the matter will not be pursued.

e.g.:

$$Z_{1t} = 1 \text{ for observations in the first group}$$
$$\phantom{Z_{1t}} = 0 \text{ for all other observations}$$

and

$$Z_{2t} = 1 \text{ for observations in the third group}$$
$$\phantom{Z_{2t}} = 0 \text{ for all other observations.}$$

If T/3 yields an integer, $\underline{W}$ will for this case look as follows, after transition to variables expressed in deviations from their mean

$$\underline{W} = \frac{1}{3}\begin{bmatrix} 2 & -1 \\ 2 & -1 \\ \vdots & \vdots \\ 2 & -1 \\ -1 & -1 \\ -1 & -1 \\ \vdots & \vdots \\ -1 & -1 \\ -1 & 2 \\ -1 & 2 \\ \vdots & \vdots \\ -1 & 2 \end{bmatrix} \tag{2.6.18}$$

The variables $\underline{W}'\underline{Y}$ and $\underline{W}'\underline{X}$ from (2.6.9) are, for the above situation, equal to:

$$\underline{W}'\underline{Y} = \frac{T}{3}\begin{bmatrix} \bar{Y}_1 - \bar{Y} \\ \bar{Y}_3 - \bar{Y} \end{bmatrix} \qquad \underline{W}'\underline{X} = \frac{T}{3}\begin{bmatrix} \bar{X}_1 - \bar{X} \\ \bar{X}_3 - \bar{X} \end{bmatrix} \tag{2.6.19}$$

where $\bar{Y}$ is the mean of all observations of Y, and $\bar{X}$ has a corresponding meaning. Simple linear regression of $\underline{W}'\underline{Y}$ on $\underline{W}'\underline{X}$ yields, as can be calculated easily from (2.6.18), the Bartlett estimator for β.

With regard to the generalization to cases with more than one regressor the following remarks can be added. Because the grouping methods of Wald and Bartlett mostly have not been recognized as special cases of the application of instrumental variables, attempts have been made to generalize the grouping principle to cases with more than one regressor. Gibson and Jowett[92] pursued the idea that regression coefficients of multiple regression are composed of simple or zero order regression coefficients. If the last-mentioned coefficients are calculated by Wald's or Bartlett's method, the multiple regression coefficients can be determined with their help. The disadvantage of this procedure is first and foremost the difficulty non-linear expressions present in establishing the variances. To illustrate the non-linearity of the estimators the regression coefficient belonging to regressor X in the regression of Z on X and Y is printed below:

$$b_{ZX.Y} = \frac{b_{ZX}-b_{YX}.b_{ZY}}{1-b_{XY}.b_{YX}} \qquad (2.6.20)$$

where b_{ZX} represents the regression coefficient from the simple regression of Z on X, b_{YX} that from the regression of Y on X, etc., each of these simple regression coefficients calculated, e.g., by Wald's method.

Hooper and Theil[93] started from the following idea. A

92. W.M. Gibson and G.H. Jowett, 'Three-group' regression analysis, parts I and II, Applied Statistics, Vol. 6 (1957), pp. 114-122 and 189-197.
93. J.W. Hooper and H. Theil, The Extension of Wald's method of fitting straight lines to multiple regression, Revue de l'Institute International de Statistique, Vol. 26 (1958), pp. 37-47.

simple linear relation is fully determined by two points, for instance the points $(\bar{Y}_2, \bar{X}_2)$ and $(\bar{Y}_1, \bar{X}_1)$. A relation between three variables is determined by 3 gravity centres, etc. We shall not elaborate their method, because - as the authors have remarked themselves - it yields less satisfactory practical results than that of Gibson and Jowett.

The difficulties met by Gibson and Jowett when they tried to generalize the grouping methods can be avoided if the grouping approach is recognized as an application of the technique of instrumental variables. For a case of two regressors a total of three dummy variables are needed in applying Wald's grouping method, viz. the constant-term variable, one binary dummy variable to represents the division into two groups of the first regressor, and another binary dummy variable to indicate the division into two groups of the second regressor. Both the estimates for the parameters and the corresponding asymptotic variances are easy to calculate with the help of, respectively, (2.6.10) and (2.6.11). It can be proved that (2.6.10) gives the same results as the method developed by Gibson and Jowett. Their approach, too, can therefore be looked upon as an application of dummy variables as instrumental variables. The three approaches to the problem of errors in variables have thus been reduced to two, namely the one applying the maximum-likelihood method and the one applying the instrumental variable method; the practical significance of the latter will now be discussed.

The drawbacks inherent to the instrumental variable method are well-known[94]:

(i) the impossibility of checking whether the

94. See J. Johnston, ibid. p. 8, p. 281.

instrumental variable, for large samples, is or is not correlated with the (non-observable) disturbance term, and

(ii) the rather high degree of arbitrariness in the choice of the instrumental variable, which may cause great fluctuations in the parameters estimated.

The first of these drawbacks should be completed by the statement that the use of dummy variables as instrumental variables as described above is very likely to introduce some degree of correlation between disturbance term and instrumental variable. Let us in this connection look once more at relation (2.6.3). The grouping methods developed by Wald and Bartlett arrange the X-observations according to size, after which the division into groups is made. With regard to instrumental variables, in this case the dummy variables, the greatest possible independence from the disturbance term in (2.6.3) should be aimed at. If the division into groups were to be made on the basis of the $\tilde{X}_t$-observations, that independence would be guaranteed, given the in most cases reasonable assumption of independence between $\tilde{X}$ and u and between $\tilde{X}$ and v. Now that the division into groups is carried out on the basis of the observed quantities $(\tilde{X} + v)$, however, correlation will be introduced - notably where groups pass into one another - between v and the instrumental variables in $\underline{W}$. This correlation is, in Wald's grouping, positive if the binary dummy variable is coded by 0 for lower, and by 1 for higher values of X.
Application of dummy variables cannot help either in overcoming the general drawbacks connected with the instrumental variable method. The introduction of correlation between disturbance term and instrumental variable, as meant above, can be remedied if the binary

dummy variables are not taken from the observed material itself but from additional information, for example in the manner to be described below.
If one wishes, for instance, to quantify an Engel curve for a certain commodity, one will collect from N families data on the quantity of the commodity used - to be denoted by $Y_i (i = 1,2,...,N)$ as well as on family income, X_i, and that as accurately as possible. Next, one will want to estimate, on the basis of the material obtained, the parameters α and β from the relation below:

$$Y_i = \alpha + \beta X_i + \varepsilon_i \qquad (2.6.20)$$

where ε_i is the disturbance in the relation relevant to the ith observation. It is not at all unlikely, however, that the income data collected show a relatively large error because the questioners frequently do not estimate correctly the real height of the family income. The situation just described is specific in that questioners are told to visit only those families whose income can be ascertained to be either below or above a certain borderline. The income level marking the upper ceiling of obligatory insurance can serve in the Netherlands as such a borderline. About each family three figures are known, viz. the relevant values of Y_i, X_i and Z_i, Z_i being zero for a family with an income below that income level, and 1 if the income is above that level. Contrary to variable X, Z does not show any observation errors. If Z is used as an instrumental variable it may reasonably be expected that the condition of independence between Z and ε is fulfilled, because the category to which a family belongs has nothing to do with the measuring errors made. Moreover it may be assumed in this situation that the third condition (see p. 79) will be met because high values of X will in most cases be accompanied by the value 1 for Z and low values of X by the value 0 for Z. If instrumental variables are based on additional information rather than on observations of the regressor, a higher-quality estimator may be expected to emerge. To what extent the use of dummy variables as instrumental variables in situations with errors in variables will lead to appreciably better practical results than, e.g., the application of OLS, remains the object of further study; the arbitrariness in the choice of instrumental variables, but always with a consistent estimator as a result, will have to be weighed up against, for instance, the inconsistent OLS estimator, which, however, has a smaller variance.

3. Dummy variables and analysis of variance and covariance

3.1. INTRODUCTION

For many econometrists analysis of variance and covariance lie at the periphery of their discipline, mainly because these methods are developed to examine the results of experiments. In econometrics experiments play a secondary role, so analysis of variance and covariance have fallen into the background. Morever in practice there is hardly ever any integration between the techniques of regression analysis and those of analysis of variance and covariance. This chapter will elaborate on that integration.
Collecting cross-section data comes nearest to an experimental situation. In section 2.3, where a classifation of individuals according to level of education and sex was given, one could have spoken of the results of an experiment, for the number of observations per cell of table 2.1 can be increased by similar observations according to one's needs. It is not surprising, therefore, that data obtained in the situation described can be analysed by regression methods as well as analysis of variance techniques. Moreover the equivalence of tests from the two methods can be demonstrated, as will become apparent in section 3.2.
Analysis of variance, a reasearch method by which the significance of a classification is examined, comprises a very wide range of models. Distinctions are made according to the number of classifications, the number of

observations per cell and the presence or absence of interaction. There is interaction when the simultaneous occurence of several characteristics has a separate effect over and above the influences of each of the individual characteristics. To keep as close as possible to the examples of chapter 2, we shall deal consecutively with analysis of variance for a simple classification (one-way analysis of variance), analysis of variance for a two-fold classification (two-way analysis of variance) and finally analysis of variance for a two-fold classification with interaction (two-way analysis of variance with interaction). To simplify the analysis of variance, the abovementioned cases will be studied for equal numbers of observations per cell.

In analysis of variance the total sum of squares of the observations is divided into a number of mutually independent subtotals. For each of the cases mentioned above there will be given below the division of the sum of squares into subtotals, the relevant analysis of variance (ANOVA) table and the test statistics taken from it.

With a simple classification into r classes with n elements per class (cell), the total sum of squares of the observations Y_{ik} ($i = 1,2,\ldots,r$ and $k = 1,\ldots n$) with mean $\bar{Y}_{..} = \frac{1}{nr} \sum_{i=1}^{r} \sum_{k=1}^{n} Y_{ik}$ is divided as follows into the variance of the cell means $\bar{Y}_{i.} = \frac{1}{n} \sum_{k=1}^{n} Y_{ik}$ and the sum of squares within the cells:

$$S_T = \sum_{i=1}^{r} \sum_{k=1}^{n} (Y_{ik}-\bar{Y}_{..})^2 = \sum_{i=1}^{r} \sum_{k=1}^{n} (Y_{ik}-\bar{Y}_{i.}-\bar{Y}_{..}+\bar{Y}_{i.})^2 =$$

$$= \sum_{i=1}^{r} \sum_{k=1}^{n} (\bar{Y}_{i.} - \bar{Y}_{..})^2 + \sum_{i=1}^{r} \sum_{k=1}^{n} (Y_{ik} - \bar{Y}_{i.})^2 = \qquad (3.1.1)$$

$$= n \sum_{i=1}^{r} (\bar{Y}_{i.} - \bar{Y}_{..})^2 + \sum_{i=1}^{r} \sum_{k=1}^{n} (Y_{ik} - \bar{Y}_{i.})^2 = S_A + S_E$$

Expression (3.1.1) is arranged in an ANOVA table[1] as follows:

Table 3.1. ANOVA table for a simple classification with n observations per cell

Source of Variation	Number of degrees of freedom	Sum of squares	Variance
Differences between the r classes	(r-1)	S_A	$S_A/(r-1) = s_A$
Differences within the r classes	(n-1)r	S_E	$S_E/(n-1)r = s_E$
Total	nr - 1	S_T	

The test statistic on the significance of the classification follows an F-distribution. It has the shape

$$F_{[(r-1),(n-1)r]} = \frac{s_A}{s_E} \qquad (3.1.2)$$

Analogously with (3.1.1), for a two-way classification with, respectively, r and s classes and n elements in each of the r.s cells, the total sum of squares is divided as follows.

1. C.R. Hicks, Fundamental concepts in the design of experiments, New York 1963, p. 27.

$$S_{T'} = \sum_{i=1}^{r} \sum_{j=1}^{s} \sum_{k=1}^{n} (Y_{ijk} - \bar{Y}_{...})^2 =$$

$$= \sum_{i=1}^{r} \sum_{j=1}^{s} \sum_{k=1}^{n} [(\bar{Y}_{i..} - \bar{Y}_{...}) + (\bar{Y}_{.j.} - \bar{Y}_{...}) +$$

$$+ (Y_{ijk} - \bar{Y}_{i..} - \bar{Y}_{.j.} + \bar{Y}_{...})]^2 =$$

$$= ns \sum_{i=1}^{r} (\bar{Y}_{i..} - \bar{Y}_{...})^2 + nr \sum_{j=1}^{s} (\bar{Y}_{.j.} - \bar{Y}_{...})^2 + \quad (3.1.3)$$

$$+ \sum_{i=1}^{r} \sum_{j=1}^{s} \sum_{k=1}^{n} (Y_{ijk} - \bar{Y}_{i..} - \bar{Y}_{.j.} + \bar{Y}_{...})^2 = S_{A'} + S_{B'} + S_{E'}$$

C.R. Hicks[2] gives as relevant ANOVA table the following set-up:

Table 3.2. ANOVA table of a two-way classification without interaction with n observations per cell

Source of Variation	Number of degrees of freedom	Sum of squares	Variance
Differences between the r classes of the first classification	(r-1)	$S_{A'}$	$S_{A'}/(r-1) = s_{A'}$
The same for the second classification	(s-1)	$S_{B'}$	$S_{B'}/(s-1) = s_{B'}$
Differences within the classes	nrs - (r+s-1)	$S_{E'}$	$S_{E'}/[(nrs-(r+s-1)] = s_{E'}$
Total	nrs - 1	$S_{T'}$	

2. Ibid. p. 92, p. 56.

The first and second classifications, respectively, are tested with the F-distributed statistics:

$$F_{[(r-1),nrs-(r+s-1)]} = \frac{s_{A'}}{s_{E'}}$$

and (3.1.4)

$$F_{[(s-1),nrs-(r+s-1)]} = \frac{s_{B'}}{s_{E'}}$$

Finally, when there is interaction between the two classifications, the total sum of squares is divided into four parts:

$$S_{T'} = \sum_{i=1}^{r} \sum_{j=1}^{s} \sum_{k=1}^{n} (Y_{ijk}-\bar{Y}_{...})^2 =$$

$$= \sum_{i=1}^{r} \sum_{j=1}^{s} \sum_{k=1}^{n} [(\bar{Y}_{i..}-\bar{Y}_{...})+(\bar{Y}_{.j.}-\bar{Y}_{...}) +$$

$$+ (\bar{Y}_{ij.}-\bar{Y}_{i..}-\bar{Y}_{.j.}+\bar{Y}_{...})+(Y_{ijk}-\bar{Y}_{ij.})]^2 =$$

$$= S_{A'}+S_{B'}+n \sum_{i=1}^{r} \sum_{j=1}^{s} (\bar{Y}_{ij.}-\bar{Y}_{i..}-\bar{Y}_{.j.}+\bar{Y}_{...})^2 +$$

$$+ \sum_{i=1}^{r} \sum_{j=1}^{s} \sum_{k=1}^{n} (Y_{ijk}-\bar{Y}_{ij.})^2 =$$

$$= S_{A'}+S_{B'}+S_{A'B'}+S_{E''} \qquad (3.1.5)$$

Arranging the data in an ANOVA table produces the followin picture[3]:

3. See C.R. Hicks, Ibid. p. 92, p. 87.

Table 3.3. ANOVA table of a two-way classification with interaction with n observations per cell

Source of Variation	Number of degrees of freedom	Sum of squares	Variance
The first classification	(r-1)	$S_{A'}$	$S_{A'}/(r-1) = s_{A'}$
The second classification	(s-1)	$S_{E'}$	$S_{B'}/(s-1) = s_{B'}$
The interaction	(r-1)(s-1)	$S_{A'B'}$	$S_{A'B'}/(r-1)(s-1) = s_{A'B'}$
Unexplained	(n-1)rs	$S_{E''}$	$S_{E''}/(n-1)rs = s_{E''}$
Total	nrs - 1	$S_{T'}$	

The significance of the interaction is tested with the F-distributed statistic:

$$F_{[(r-1)(s-1),(n-1)rs]} = \frac{s_{A'B'}}{s_{E''}} \qquad (3.1.6)$$

In section 3.2. it will be demonstrated that the test statistic of (3.1.2) corresponds with the one to test the significance of R^2, relevant to the regression of Y_{ik} on the constant-term variable and (r-1) binary dummy variables. For the test statistics of (3.1.4) and (3.1.6) the corresponding tests will be derived from the regression analysis for the example in section 2.3, where the classifications consist of three and two classes, respectively.

Analysis of covariance combines, as Malinvaud[4] points out, regression analysis with analysis of variance. One

---

4. Ibid. p. 3, p. 233.

can indeed imagine a situation in which a dependent variable is affected by one or more real variables as well as by one or more classifications. Cochran[5] mentions as a reason for applying analysis of covariance the elimination of the effect of disturbing real variables from the quantity to be observed. If, for example, one wishes to study income differences with the help of classifications according to level of education and sex, then the age of the persons involved cannot be left out of account. The income differences must be corrected for the influence of the last-mentioned variable, often called co-variable, before the influences of the level of schooling and sex are investigated. A second possibility for using analysis of covariance Cochran finds in testing the equality of parameters in linear relations which each are valid separately for one class of a classification. The latter possibility will be discussed in connection with Chow's test[6] in chapter 4, so that it can be omitted from section 3.3, where analysis of covariance will be dealt with.

For the significance test to be applied to a simple classification into r classes with n observations per cell, use is made of regression coefficients $\hat{\beta}$ and $\hat{\beta}_*$, respectively, from the regression of regressand Y_{ik}[7] on co-variable X_{ik}, and from the regression of $(Y_{ik}-\bar{Y}_{i.})$ on $(X_{ik}-\bar{X}_{i.})$. In Cochran[8] a test statistic to test the significance of the classification is found, which can be written as:

---

5. W.G. Cochran, Analysis of covariance: its nature and uses, Biometrics, Vol. 13, September 1957, pp. 261-281.
6. Ibid. p. 11.
7. The kth observation in the ith class is indicated by the indices i and k.
8. See footnote number 5, p. 269.

$$F[(r-1),(nr-r-1)] = \frac{(nr-r-1)}{r-1} \frac{\left[n \sum_{i=1}^{r} (\bar{Y}_{i.}-\bar{Y}_{..})^2 - \hat{\beta}^2 \sum_{i=1}^{r} \sum_{k=1}^{n} (X_{ik}-\bar{X}_{..})^2 + \hat{\beta}_*^2 \sum_{i=1}^{r} \sum_{k=1}^{n} (X_{ik}-\bar{X}_{i.})^2\right]}{\left[\sum_{i=1}^{r} \sum_{k=1}^{n} (Y_{ik}-\bar{Y}_{i.})^2 - \hat{\beta}_*^2 \sum_{i=1}^{r} \sum_{k=1}^{r} (X_{ik}-\bar{X}_{i.})^2\right]} \qquad (3.1.7)$$

Section 3.3. contains the proof that the test statistic (3.1.7) corresponds to a test from regression analysis.

3.2. A COMPARISON BETWEEN ANALYSIS OF VARIANCE AND REGRESSION ANALYSIS WITH DUMMY VARIABLES AS REGRESSORS

Below it will be proved that three much-used significance tests[9] from regression analysis[10], if applied with dummy variables as regressors, correspond with tests from analysis of variance. The argument concerns notably the equivalence of the test on all parameters of a linear relation, or on the square of the relevant multiple-correlation coefficient R, with the test from (3.1.2) on p. 92 (case A), the t-test on one parameter with the test from (3.1.6) on p. 95 (case B), and of the test on more than one parameter with the test from (3.1.4) on p. 94 (case C). Cases B and C will be treated by means of an example from section 2.3, in which the variables 'level of education' and 'sex' are operated. This example has been chosen, on the one hand because the exposition can be simple, since both magnitudes are subdivided into a small number of classes - three for the level of

---

9. To test the significance of a parameter, it is checked whether it differs significantly from zero.
10. The regression model with the relevant assumptions is found on p. 18.

education and two for sex - , and on the other hand because case B, as will become clear, is relevant only when there is a classification of exactly two classes. For case A no appreciable simplification in the exposition can be achieved by choosing a qualitative magnitude that contains only a few classes; so in case A an unspecified number of classes, indicated by r, will be used.

Both Malinvaud[11] and Kendall and Stuart[12], when comparing regression analysis with analyse of variance, use dummy variables expressed in deviations from the relevant means. That approach will not be followed in this section, for one thing because the transformation of dummy variables, mentioned above, is not necessary for a comparison of regression analysis with analysis of variance, and for another because we want to build on the results from section 2.3, in which the dummy variables are composed of ones and zeros.

The tests of analysis of variance represented in section 3.1 all refer to cases with an equal number of observations, viz. n, per cell. Analysis of variance with unequal numbers of observations per cell are less frequent in practice and can, therefore, be left out of consideration. For a meaningful comparison between analysis of variance and regression analysis, all classes to be used in the present section will hold the same number of elements.

Case A.

Before proof can be given of the equivalence of the significance test on R^2 and the test statistic of

11. Ibid. p. 3, pp. 229 ff.
12. Ibid. p. 8, pp. 14 and 15.

(3.1.2) on page 92, the linear relation for which R^2 is to be calculated will have to be set up. This linear relation indicates the relationship between the variable 'to be explained', Y_a[13] and the 'explanatory' qualitative magnitude. Within the qualitative magnitude, analogously to (3.1.2) r categories with n elements each are distinguished. In a regression r categories can be expressed by r regressors, viz. the constant-term variable and (r-1) binary dummy variables, see p. 46. If a vector of nr observations of variable Y_a is denoted by $\underline{Y}_a$, the dummy variables by $\underline{1}_a$, $\underline{Z}_{a2}$, $\underline{Z}_{a3}$,...,$\underline{Z}_{ar}$, the parameters by, respectively β_{a1}, β_{a2},...,β_{ar}, and the disturbances by $\underline{\varepsilon}_a$, the linear relation can be written as:

$$\underline{Y}_a = \underline{1}_a\beta_{a1} + \underline{Z}_{a2}\beta_{a2} + \underline{Z}_{a3}\beta_{a3} + \ldots + \underline{Z}_{ar}\beta_{ar} + \underline{\varepsilon}_a \tag{3.2.1}$$

If the kth observation (k = 1,2,...,n) in the ith class (i = 1,2,...,r) is indicated for the variable Y_a by Y_{aik} and for the disturbance term by ε_{aik}, then (3.2.1) can be written in full as:

13. All variables and coefficients are given an index corresponding with the letter of the case concerned. To make the comparison meaningful, the regression analysis refers to the same data as the analysis of variance.

$$
\begin{bmatrix}
Y_{a11}\\ Y_{a12}\\ \cdot\\ \cdot\\ \cdot\\ Y_{a1n}\\ Y_{a21}\\ Y_{a22}\\ \cdot\\ \cdot\\ \cdot\\ Y_{a2n}\\ \cdot\\ \cdot\\ \cdot\\ \cdot\\ \cdot\\ \cdot\\ \cdot\\ \cdot\\ \cdot\\ \cdot\\ Y_{ar1}\\ Y_{ar2}\\ \cdot\\ \cdot\\ \cdot\\ Y_{arn}
\end{bmatrix}
=
\begin{bmatrix}
1 & 1 & 0 & & & & 0\\
1 & 1 & 0 & & & & 0\\
\cdot & \cdot & \cdot & & & & \cdot\\
\cdot & \cdot & \cdot & & & & \cdot\\
\cdot & \cdot & \cdot & & & & \cdot\\
\cdot & 1 & 0 & & & & \cdot\\
\cdot & 0 & 1 & & & & \cdot\\
\cdot & 0 & 1 & & & & \cdot\\
\cdot & \cdot & \cdot & & & & \cdot\\
\cdot & \cdot & \cdot & & & & \cdot\\
\cdot & \cdot & \cdot & & & & \cdot\\
\cdot & \cdot & 1 & & & & \cdot\\
\cdot & \cdot & 0 & & & & \cdot\\
\cdot & \cdot & 0 & \cdot & \cdot & \cdot & \cdot\\
\cdot & \cdot & \cdot & & & & \cdot\\
\cdot & \cdot & \cdot & & & & \cdot\\
\cdot & \cdot & \cdot & & & & 0\\
\cdot & \cdot & \cdot & & & & 1\\
\cdot & \cdot & \cdot & & & & 1\\
\cdot & \cdot & \cdot & & & & \cdot\\
\cdot & \cdot & \cdot & & & & \cdot\\
\cdot & \cdot & \cdot & & & & 1\\
\cdot & \cdot & \cdot & & & & 0\\
\cdot & \cdot & \cdot & & & & 0\\
\cdot & \cdot & \cdot & & & & \cdot\\
\cdot & \cdot & \cdot & & & & \cdot\\
\cdot & \cdot & \cdot & & & & \cdot\\
1 & 0 & 0 & & & & 0
\end{bmatrix}
\begin{bmatrix}
\beta_{a1}\\ \beta_{a2}\\ \cdot\\ \cdot\\ \cdot\\ \beta_{ar}
\end{bmatrix}
+
\begin{bmatrix}
\varepsilon_{a11}\\ \varepsilon_{a12}\\ \cdot\\ \cdot\\ \cdot\\ \varepsilon_{a1n}\\ \varepsilon_{a21}\\ \varepsilon_{a22}\\ \cdot\\ \cdot\\ \cdot\\ \varepsilon_{a2n}\\ \cdot\\ \cdot\\ \cdot\\ \cdot\\ \cdot\\ \cdot\\ \cdot\\ \cdot\\ \cdot\\ \cdot\\ \varepsilon_{ar1}\\ \varepsilon_{ar2}\\ \cdot\\ \cdot\\ \cdot\\ \varepsilon_{arn}
\end{bmatrix}
\qquad (3.2.2)
$$

in which the first column of the matrix is equal to $\underline{1}_a$, the second to $\underline{Z}_{a2}$, the third to $\underline{Z}_{a3}$, and so on.
The relation in (3.2.2) can be written much more concisely by indicating the nr × 1 matrix by $\underline{D}_a$ and the vector

of parameters by $\underline{\beta}_a$, thus:

$$\underline{Y}_a = \underline{D}_a\underline{\beta}_a + \underline{\varepsilon}_a \qquad (3.2.3)$$

The least-squares estimator for $\underline{\beta}_a$, which is required for the calculation of R^2, reads according to formula (2.1.7) on p. 19, $\underline{Y}$ being substituted by $\underline{Y}_a$, $\underline{X}$ by $\underline{D}_a$ and $\hat{\underline{\beta}}$ by $\hat{\underline{\beta}}_a$:

$$\hat{\underline{\beta}}_a = (\underline{D}_a'\underline{D}_a)^{-1}\underline{D}_a'\underline{Y}_a \qquad (3.2.4)$$

Written out in its composing elements, (3.2.4) has the following form:

$$\begin{bmatrix} \hat{\beta}_{a1} \\ \hat{\beta}_{a2} \\ \hat{\beta}_{a3} \\ \cdot \\ \cdot \\ \cdot \\ \hat{\beta}_{ar} \end{bmatrix} = \begin{bmatrix} nr & n & n & . & . & . & n \\ n & n & 0 & . & . & . & 0 \\ n & 0 & n & . & . & . & 0 \\ \cdot & \cdot & \cdot & \cdot & & & \cdot \\ \cdot & \cdot & \cdot & & \cdot & & \cdot \\ \cdot & \cdot & \cdot & & & \cdot & \cdot \\ n & 0 & 0 & . & . & . & n \end{bmatrix}^{-1} \begin{bmatrix} \sum_{i=1}^{r}\sum_{k=1}^{n} Y_{aik} \\ \sum_{k=1}^{n} Y_{a1k} \\ \sum_{k=1}^{n} Y_{a2k} \\ \cdot \\ \cdot \\ \cdot \\ \sum_{k=1}^{n} Y_{a(r-1)k} \end{bmatrix} \qquad (3.2.5)$$

If the above inverse is calculated, and the mean of the observations of variable Y_a from the ith class is indicated by $\bar{Y}_{ai.}$ and that of all observations by $\bar{Y}_{a..}$, (3.2.5) can be written as:

$$\begin{bmatrix} \hat{\beta}_{a1} \\ \hat{\beta}_{a2} \\ \hat{\beta}_{a3} \\ \cdot \\ \cdot \\ \cdot \\ \hat{\beta}_{ar} \end{bmatrix} = \frac{1}{n} \begin{bmatrix} 1 & -1 & -1 & \cdot & \cdot & \cdot & -1 \\ -1 & 2 & +1 & \cdot & \cdot & \cdot & +1 \\ -1 & +1 & 2 & \cdot & \cdot & \cdot & +1 \\ \cdot & \cdot & \cdot & \cdot & & & \cdot \\ \cdot & \cdot & \cdot & & \cdot & & \cdot \\ \cdot & \cdot & \cdot & & & \cdot & \cdot \\ -1 & +1 & +1 & \cdot & \cdot & \cdot & 2 \end{bmatrix} \begin{bmatrix} nr\bar{Y}_{a..} \\ n\,\bar{Y}_{a1.} \\ n\,\bar{Y}_{a2.} \\ \cdot \\ \cdot \\ \cdot \\ n\,\bar{Y}_{a(r-1).} \end{bmatrix} =$$

$$= \begin{bmatrix} r\,\bar{Y}_{a..} - \bar{Y}_{a1.} - \bar{Y}_{a2.} - \dots - \bar{Y}_{a(r-1).} \\ -r\,\bar{Y}_{a..} + 2\bar{Y}_{a1.} + \bar{Y}_{a2.} + \dots + \bar{Y}_{a(r-1).} \\ -r\,\bar{Y}_{a..} + \bar{Y}_{a1.} + 2\bar{Y}_{a2.} + \dots + \bar{Y}_{a(r-1).} \\ \cdot \\ \cdot \\ \cdot \\ -r\,\bar{Y}_{a..} + \bar{Y}_{a1.} + \bar{Y}_{a2.} + \dots + 2\bar{Y}_{a(r-1).} \end{bmatrix} \qquad (3.2.6)$$

Expression (3.2.6) can be simplified, by means of the equality $r\bar{Y}_{a..} = \sum_{i=1}^{r} \bar{Y}_{ai.}$, to

$$\begin{bmatrix} \hat{\beta}_{a1} \\ \hat{\beta}_{a2} \\ \hat{\beta}_{a3} \\ \cdot \\ \cdot \\ \cdot \\ \hat{\beta}_{ar} \end{bmatrix} = \begin{bmatrix} \bar{Y}_{ar.} \\ \bar{Y}_{a1.} - \bar{Y}_{ar.} \\ \bar{Y}_{a2.} - \bar{Y}_{ar.} \\ \cdot \\ \cdot \\ \cdot \\ \bar{Y}_{a(r-1).} - \bar{Y}_{ar.} \end{bmatrix} \qquad (3.2.7)$$

The result in (3.2.7) shows that the estimator for the constant in the regression of $\underline{Y}_a$ on $\underline{D}_a$ is equal to the mean of the Y_a-values of the rth class. The estimated parameters for the binary dummy variables are equal to, respectively, the differences between the means of variable Y_a over the n observations of the first up to and including the $(r-1)^{th}$ class and the mean of the rth class.

Classification of the observations of Y_a over r classes is meaningful only if within the population the means of the classes differ. The null hypothesis of the test on the significance of the classiifcation will read, therefore[14]:

$$\bar{Y}_{a1.} = \bar{Y}_{a2.} = \ldots = \bar{Y}_{ar.}$$

which can also be written as:

$$\bar{Y}_{a1.} - \bar{Y}_{ar.} = \bar{Y}_{a2.} - \bar{Y}_{ar.} = \ldots = \bar{Y}_{a(r-1).} - \bar{Y}_{ar.} = 0 \qquad (3.2.8)$$

Testing the null hypothesis as expressed in (3.2.8) corresponds, in terms of the parameters in (3.2.1), with testing the hypothesis:

$$\beta_{a2} = \beta_{a3} = \ldots = \beta_{ar} = 0 \qquad (3.2.9)$$

A simultaneous test on all parameters from a regression with the exception of the constant is frequently carried out with the test statistic of the significance of R^2 [15] given below:

$$F_{[(r-1),(nr-r)]} = \frac{(nr-r)}{(r-1)} \frac{R^2}{1-R^2} \qquad (3.2.10)$$

As appears from the formula for R^2 in (2.1.9) on page 20, for the determination of R^2 the estimated disturbances

---

14. What follows refers to population parameters; for convenience sake no new symbols have been introduced.
15. See the test statistic (2.1.15) on page 23.

from the regression involved are required. The estimated disturbances of the regression of $\underline{Y}_a$ on $\underline{D}_a$ are obtained by means of (2.1.8) on page 19, with substitution of $\underline{X}$ by $\underline{D}_a$, $\underline{Y}$ by $\underline{Y}_a$ and $\hat{\underline{\beta}}$ by $\hat{\underline{\beta}}_a$,

$$\hat{\underline{\varepsilon}}_a = \underline{Y}_a - \underline{D}_a\hat{\underline{\beta}}_a \tag{3.2.11}$$

If $\hat{\underline{\beta}}_a$ from (3.2.7) is then substituted into (3.2.11), it is easy to prove that the i,kth element of $\hat{\underline{\varepsilon}}_a$ looks as follows:

$$\hat{\varepsilon}_{aik} = Y_{aik} - \bar{Y}_{ai.} \tag{3.2.12}$$

Substitution of the elements of $\hat{\underline{\varepsilon}}$ and $\underline{Y}_a$ into the formula of R^2 from (2.1.9) results in:

$$R^2 = 1 - \frac{\sum_{i=1}^{r}\sum_{k=1}^{n}(Y_{aik}-\bar{Y}_{ai.})^2}{\sum_{i=1}^{r}\sum_{k=1}^{n}(Y_{aik}-\bar{Y}_{a..})^2} \tag{3.2.13}$$

Finally, substitution into (3.2.10) of R^2 from (3.2.13) yields the following test statistic

$$F_{[(r-1),(nr-r)]} =$$

$$= \frac{(nr-r)}{(r-1)} \frac{\sum_{i=1}^{r}\sum_{k=1}^{n}(Y_{aik}-\bar{Y}_{a..})^2 - \sum_{i=1}^{r}\sum_{k=1}^{n}(Y_{aik}-Y_{ai.})^2}{\sum_{i=1}^{r}\sum_{k=1}^{n}(Y_{aik}-\bar{Y}_{ai.})^2} \tag{3.2.14}$$

The numerator of (3.2.14) can be simplified into $n\sum_{i=1}^{r}(\bar{Y}_{ai.}-\bar{Y}_{a..})^2$.

In this form, the test statistic on the significance of the parameters $\beta_{a2}, \beta_{a3},\ldots,\beta_{ar}$ is - apart from the index a indicating that case A is concerned - identical with test statistic (3.1.2) from the analysis of variance on p. 92 which serves to test the significance of a classification into r classes. The equivalence of the two tests involved has thus been proved.

Case B

The t-test on the significance of one parameter (see (2.1.18) on page 24) and the test on the significance of a classification with s classes (see (3.1.4) second line) will correspond only if s = 2, for in that case the classification can be represented by one binary dummy variable. The influence of the relevant dummy variable is indicated by the corresponding parameter; because there is only one parameter the t-test can serve to check its significance. In this situation it becomes possible to compare the t-test with the corresponding test in analysis of variance. Because in the example from section 2.3 on p. 58 the classification according to sex consists of two categories, this example will be used for the comparison of the abovementioned tests.
The test statistic on the significance of the second classification from (3.1.4), for which in this case s = 2, has been derived on the assumption that no interaction occurs between the two classifications[16]. The representation of the qualitative magnitude 'level of education' by three classes and of the qualitative magnitude 'sex' by two classes in a regression leads, in the absence of interaction, to four dummy variables (see

16. In case C an example will be taken in which there is interaction.

(2.3.20) on page 58). Apart from the constant-term variable $\underline{1}_b$[17], two binary dummy variables are required for the magnitude 'level of education' and one binary dummy variable for 'sex' to be denoted by $\underline{Z}_{b2}$, $\underline{Z}_{b3}$ and $\underline{Z}_{b4}$, respectively. The linear relation with as regressand $\underline{Y}_b$ and as parameters β_{b1}, β_{b2}, β_{b3}, and β_{b4}, reads:

$$\underline{Y}_b = \underline{1}_b\beta_{b1} + \underline{Z}_{b2}\beta_{b2} + \underline{Z}_{b3}\beta_{b3} + \underline{Z}_{b4}\beta_{b4} + \underline{\varepsilon}_b \qquad (3.2.15)$$

If the kth observation ($k = 1,2,\ldots,n$) in the ith class of the first classification ($i = 1,2,3$) and in the jth class of the second classification ($j = 1,2$) is indicated, for variable Y_b and the disturbance term, by, respectively, Y_{bijk} and ε_{bijk}, then (3.2.15), written in full, will become:

17. The symbols used for case B have had an index b added to them.

$$
\begin{bmatrix}
Y_{b111} \\ Y_{b112} \\ \cdot \\ \cdot \\ Y_{b11n} \\ Y_{b121} \\ Y_{b122} \\ \cdot \\ \cdot \\ Y_{b12n} \\ Y_{b211} \\ Y_{b212} \\ \cdot \\ \cdot \\ Y_{b21n} \\ Y_{b221} \\ Y_{b222} \\ \cdot \\ \cdot \\ Y_{b22n} \\ Y_{b311} \\ Y_{b312} \\ \cdot \\ \cdot \\ Y_{b31n} \\ Y_{b321} \\ Y_{b322} \\ \cdot \\ \cdot \\ Y_{b32n}
\end{bmatrix}
=
\begin{bmatrix}
1 & 1 & 0 & 1 \\
1 & 1 & 0 & 1 \\
\cdot & \cdot & \cdot & \cdot \\
\cdot & \cdot & \cdot & \cdot \\
\cdot & \cdot & \cdot & 1 \\
\cdot & \cdot & \cdot & 0 \\
\cdot & \cdot & \cdot & 0 \\
\cdot & \cdot & \cdot & \cdot \\
\cdot & \cdot & \cdot & \cdot \\
\cdot & 1 & 0 & 0 \\
\cdot & 0 & 1 & 1 \\
\cdot & 0 & 1 & 1 \\
\cdot & \cdot & \cdot & \cdot \\
\cdot & \cdot & \cdot & \cdot \\
\cdot & \cdot & \cdot & 1 \\
\cdot & \cdot & \cdot & 0 \\
\cdot & \cdot & \cdot & 0 \\
\cdot & \cdot & \cdot & \cdot \\
\cdot & \cdot & \cdot & \cdot \\
\cdot & \cdot & 1 & 0 \\
\cdot & \cdot & 0 & 1 \\
\cdot & \cdot & 0 & 1 \\
\cdot & \cdot & \cdot & \cdot \\
\cdot & \cdot & \cdot & \cdot \\
\cdot & \cdot & \cdot & 1 \\
\cdot & \cdot & \cdot & 0 \\
\cdot & \cdot & \cdot & 0 \\
\cdot & \cdot & \cdot & \cdot \\
\cdot & \cdot & \cdot & \cdot \\
1 & 0 & 0 & 0
\end{bmatrix}
\begin{bmatrix}
\beta_{b1} \\ \beta_{b2} \\ \beta_{b3} \\ \beta_{b4}
\end{bmatrix}
+
\begin{bmatrix}
\varepsilon_{b111} \\ \varepsilon_{b112} \\ \cdot \\ \cdot \\ \varepsilon_{b11n} \\ \varepsilon_{b121} \\ \varepsilon_{b122} \\ \cdot \\ \cdot \\ \varepsilon_{b12n} \\ \varepsilon_{b211} \\ \varepsilon_{b212} \\ \cdot \\ \cdot \\ \varepsilon_{b21n} \\ \varepsilon_{b221} \\ \varepsilon_{b222} \\ \cdot \\ \cdot \\ \varepsilon_{b22n} \\ \varepsilon_{b311} \\ \varepsilon_{b312} \\ \cdot \\ \cdot \\ \varepsilon_{b31n} \\ \varepsilon_{b321} \\ \varepsilon_{b322} \\ \cdot \\ \cdot \\ \varepsilon_{b32n}
\end{bmatrix}
\qquad (3.2.16)
$$

in which $\underline{1}_b$ is the first column of the matrix, $\underline{Z}_{b2}$ the second, etc. The only difference between the matrix in (3.2.16) and matrix $\underline{W}_3$, printed on p. 58 and referring to the same example, lies in the number of elements per class, which is n in (3.2.16), and changed from one class to another in $\underline{W}_3$ (see table 2.1 on page 59).

By a calculation analogous with that of the least-squares estimator for $\hat{\underline{\beta}}$ in (3.2.4) one obtains for relation (3.2.15):

$$\begin{bmatrix} \hat{\underline{\beta}}_{b1} \\ \hat{\underline{\beta}}_{b2} \\ \hat{\underline{\beta}}_{b3} \\ \hat{\underline{\beta}}_{b4} \end{bmatrix} = \begin{bmatrix} 6n & 2n & 2n & 3n \\ 2n & 2n & 0 & n \\ 2n & 0 & 2n & n \\ 3n & n & n & 3n \end{bmatrix}^{-1} \begin{bmatrix} \sum_{i=1}^{3} \sum_{j=1}^{2} \sum_{k=1}^{n} Y_{bijk} \\ \sum_{j=1}^{2} \sum_{k=1}^{n} Y_{b1jk} \\ \sum_{j=1}^{2} \sum_{k=1}^{n} Y_{b2jk} \\ \sum_{i=1}^{3} \sum_{k=1}^{n} Y_{bi1k} \end{bmatrix} \qquad (3.2.17)$$

If the mean calculated over the observations of variable Y_b in the ith class of the first classification is indicated by $\bar{Y}_{bi..}$ and similarly the mean of the jth class of the second classification by $\bar{Y}_{b.j.}$ and the mean over all observations by $\bar{Y}_{b...}$, (3.2.17) may be rewritten, after inversion of the matrix, as:

$$\begin{bmatrix} \hat{\beta}_{b1} \\ \hat{\beta}_{b2} \\ \hat{\beta}_{b3} \\ \hat{\beta}_{b4} \end{bmatrix} = \frac{1}{n} \begin{bmatrix} \frac{2}{3} & -\frac{1}{2} & -\frac{1}{2} & -\frac{1}{3} \\ -\frac{1}{2} & 1 & \frac{1}{2} & 0 \\ -\frac{1}{2} & \frac{1}{2} & 1 & 0 \\ -\frac{1}{3} & 0 & 0 & \frac{2}{3} \end{bmatrix} \begin{bmatrix} 6n\ \bar{Y}_{b...} \\ 2n\ \bar{Y}_{b1..} \\ 2n\ \bar{Y}_{b2..} \\ 3n\ \bar{Y}_{b.1.} \end{bmatrix} =$$

$$= \begin{bmatrix} \bar{Y}_{b3..} + \bar{Y}_{b.2.} - \bar{Y}_{b...} \\ \bar{Y}_{b1..} - \bar{Y}_{b3..} \\ \bar{Y}_{b2..} - \bar{Y}_{b3..} \\ \bar{Y}_{b.1.} - \bar{Y}_{b.2.} \end{bmatrix} \qquad (3.2.18)$$

Parameter β_{b4} represents in (3.2.15) the influence of the binary dummy variable $\underline{Z}_{b4}$, $\underline{Z}_{b4}$ standing for the qualitative magnitude 'sex'. Classification according to sex is meaningful only if there is a difference between the means calculated over the observations of variable Y_b corresponding to men and women, respectively. The relevant null hypothesis reads (for the population) $\bar{Y}_{b.1.} = \bar{Y}_{b.2.}$; in terms of the parameter β_{b4} this means testing its significance.

The significance of a parameter is tested with the help of the t-distributed test statistic (2.1.18) see page 24. The symbols in (2.1.18) should in this case be: $\hat{\beta}_i = \hat{\beta}_{b4}$, and $c_{ii} = (4,4)^{th}$ element from the covariance matrix of $\hat{\underline{\beta}}_b$, see (3.2.18). For the standard deviation $\hat{\sigma}$ one needs, as appears from (2.1.8) on page 19, the estimated disturbances. The estimated disturbances of the relation (3.2.15) are obtained by filling in the estimated β_b's into (3.2.16). For $\hat{\varepsilon}_{bijk}$ there results:

$$\hat{\varepsilon}_{bijk} = Y_{bijk} - \bar{Y}_{bi..} - \bar{Y}_{b.j.} + Y_{b...} \qquad (3.2.19)$$

It follows for $\hat{\sigma}$ from (2.1.8), (in this example T = 6n and Λ = 4):

$$\hat{\sigma} = \sqrt{\frac{1}{(6n-4)} \sum_{i=1}^{3} \sum_{j=1}^{2} \sum_{k=1}^{n} (Y_{bijk} - \bar{Y}_{bi..} - \bar{Y}_{b.j.} + \bar{Y}_{b...})^2} \tag{3.2.20}$$

Substitution of the above terms into (2.1.10) yields the t-distributed test statistic set out below for the test on the significance of β_{b4}

$$t_{[6n-4]} =$$

$$= \frac{(\bar{Y}_{b.1.} - \bar{Y}_{b.2.})}{\sqrt{\frac{1}{6n-4} \sum_{i=1}^{3} \sum_{j=1}^{2} \sum_{k=1}^{n} (Y_{bijk} - \bar{Y}_{bi..} - \bar{Y}_{b.j.} + \bar{Y}_{b...})^2} \sqrt{\frac{2}{3} \cdot \frac{1}{n}}} \tag{3.2.21}$$

The square of a $t_{[6n-4]}$-distributed variable follows an $F_{[1,(6n-4)]}$-distribution[18]; so (3.2.21) can be transformed into

$$F_{[1,(6n-4)]} =$$

$$= \frac{(6n-4)}{2} \frac{3n(\bar{Y}_{b.1.} - \bar{Y}_{b.2.})^2}{\sum_{i=1}^{3} \sum_{j=1}^{2} \sum_{k=1}^{n} (Y_{bijk} - \bar{Y}_{bi..} - \bar{Y}_{b.j.} + \bar{Y}_{b...})^2} \tag{3.2.22}$$

Apart from index b, test statistic (3.2.22) is identical with the one from the lower line of (3.1.4), after r = 3 and s = 2 have been filled in there, and after in (3.2.22) the expression $\frac{1}{2}(\bar{Y}_{b.1.} - \bar{Y}_{b.2.})^2$ has been rewritten

18. M.G. Kendall and A. Stuart, The advanced theory of statistics, Vol. 1, London 1958, p. 382.

as $\sum_{j=1}^{2}(\bar{Y}_{b.j.}-\bar{Y}_{b...})^2$. Thus it has been proved that the t-test on the significance of the parameter of the binary dummy variable, indicating in a regression with dummy variables as regressors the influence of a qualitative magnitude with two categories, is equivalent with the corresponding significance test from the analysis of variance on the relevant classification.

The significance of the first classification can be verified in this example with the help of the simultaneous significance test on β_{b_2} and β_{b_3}. The test statistic used for it will be discussed under case C. When this testing is carried out, a test statistic is obtained which is equal to the corresponding test statistic from the analysis of variance on the significance of the first classification. The proof of this statement is not elaborated here; it is broadly analogous to the argument of case C.

Case C

To draw a comparison between the significance test on interaction of two classifications, see (3.1.6), and a test from regression analysis, one needs a regression with dummy variables in which a number of regressors represent the interaction. The example of the magnitudes 'level of education' and 'sex' with interaction (see pp. 60-61) can serve for the purpose. The matrix of the regressors is written out on p. 60, indicated by $\underline{W}_4$[19]. If the columns in this matrix are indicated, consecutively, by $\underline{1}_c$[20] for the constant-term variable, $\underline{Z}_{c2}$ and $\underline{Z}_{c3}$ for

---

19. As in the cases A and B each cell from now on contains n observations.
20. All symbols used in Case C carry the index c.

the classification according to level of education, $\underline{Z}_{c4}$ for the classification according to sex, and $\underline{Z}_{c5}$ and $\underline{Z}_{c6}$ for the interaction between the two classifications, the linear relation with variable $\underline{Y}_c$ as regressand can be written as follows:

$$\underline{Y}_c = \underline{1}_c \beta_{c1} + \underline{Z}_{c2} \beta_{c2} + \underline{Z}_{c3} \beta_{c3} + \underline{Z}_{c4} \beta_{c4} + \underline{Z}_{c5} \beta_{c5} + \\ + \underline{Z}_{c6} \beta_{c6} + \underline{\varepsilon}_c \qquad (3.2.23)$$

The parameters of the linear relation have been indicated by β_c's and the disturbances by $\underline{\varepsilon}_c$. Each observation of variable Y_c and of the disturbance is provided with three indices beside index c, viz. i for the relevant class of the (first) classification according to level of education (i = 1,2,3), j for the class of the (second) classification according to sex (j = 1,2) and k for the number of an observation within a cell (k = 1,2,...,n); the general indications of the observations of Y_c and ε_c read, respectively, Y_{cijk} and ε_{cijk}. Written in full (3.2.23) looks as follows:

$$
\begin{bmatrix}
Y_{C111} \\ Y_{C112} \\ \cdot \\ \cdot \\ \cdot \\ Y_{C11n} \\ Y_{C121} \\ Y_{C122} \\ \cdot \\ \cdot \\ \cdot \\ Y_{C12n} \\ Y_{C211} \\ Y_{C212} \\ \cdot \\ \cdot \\ \cdot \\ Y_{C21n} \\ Y_{C221} \\ Y_{C222} \\ \cdot \\ \cdot \\ \cdot \\ Y_{C22n} \\ Y_{C311} \\ Y_{C312} \\ \cdot \\ \cdot \\ \cdot \\ Y_{C31n} \\ Y_{C321} \\ Y_{C322} \\ \cdot \\ \cdot \\ \cdot \\ Y_{C32n}
\end{bmatrix}
=
\begin{bmatrix}
1 & 1 & 0 & 1 & 1 & 0 \\
1 & 1 & 0 & 1 & 1 & 0 \\
\cdot & \cdot & \cdot & \cdot & \cdot & \cdot \\
\cdot & \cdot & \cdot & \cdot & \cdot & \cdot \\
\cdot & \cdot & \cdot & \cdot & \cdot & \cdot \\
\cdot & \cdot & \cdot & 1 & 1 & \cdot \\
\cdot & \cdot & \cdot & 0 & 0 & \cdot \\
\cdot & \cdot & \cdot & 0 & 0 & \cdot \\
\cdot & \cdot & \cdot & \cdot & \cdot & \cdot \\
\cdot & \cdot & \cdot & \cdot & \cdot & \cdot \\
\cdot & \cdot & \cdot & \cdot & \cdot & \cdot \\
\cdot & 1 & 0 & 0 & \cdot & 0 \\
\cdot & 0 & 1 & 1 & \cdot & 1 \\
\cdot & 0 & 1 & 1 & \cdot & 1 \\
\cdot & \cdot & \cdot & \cdot & \cdot & \cdot \\
\cdot & \cdot & \cdot & \cdot & \cdot & \cdot \\
\cdot & \cdot & \cdot & \cdot & \cdot & \cdot \\
\cdot & \cdot & \cdot & 1 & \cdot & 1 \\
\cdot & \cdot & \cdot & 0 & \cdot & 0 \\
\cdot & \cdot & \cdot & 0 & \cdot & 0 \\
\cdot & \cdot & \cdot & \cdot & \cdot & \cdot \\
\cdot & \cdot & \cdot & \cdot & \cdot & \cdot \\
\cdot & \cdot & \cdot & \cdot & \cdot & \cdot \\
\cdot & \cdot & 1 & 0 & \cdot & \cdot \\
\cdot & \cdot & 0 & 1 & \cdot & \cdot \\
\cdot & \cdot & 0 & 1 & \cdot & \cdot \\
\cdot & \cdot & \cdot & \cdot & \cdot & \cdot \\
\cdot & \cdot & \cdot & \cdot & \cdot & \cdot \\
\cdot & \cdot & \cdot & \cdot & \cdot & \cdot \\
\cdot & \cdot & \cdot & 1 & \cdot & \cdot \\
\cdot & \cdot & \cdot & 0 & \cdot & \cdot \\
\cdot & \cdot & \cdot & 0 & \cdot & \cdot \\
\cdot & \cdot & \cdot & \cdot & \cdot & \cdot \\
\cdot & \cdot & \cdot & \cdot & \cdot & \cdot \\
\cdot & \cdot & \cdot & \cdot & \cdot & \cdot \\
1 & 0 & 0 & 0 & 0 & 0
\end{bmatrix}
\begin{bmatrix}
\beta_{C1} \\ \beta_{C2} \\ \beta_{C3} \\ \beta_{C4} \\ \beta_{C5} \\ \beta_{C6}
\end{bmatrix}
+
\begin{bmatrix}
\varepsilon_{C111} \\ \varepsilon_{C112} \\ \cdot \\ \cdot \\ \cdot \\ \varepsilon_{C11n} \\ \varepsilon_{C121} \\ \varepsilon_{C122} \\ \cdot \\ \cdot \\ \cdot \\ \varepsilon_{C12n} \\ \varepsilon_{C211} \\ \varepsilon_{C212} \\ \cdot \\ \cdot \\ \cdot \\ \varepsilon_{C21n} \\ \varepsilon_{C221} \\ \varepsilon_{C222} \\ \cdot \\ \cdot \\ \cdot \\ \varepsilon_{C22n} \\ \varepsilon_{C311} \\ \varepsilon_{C312} \\ \cdot \\ \cdot \\ \cdot \\ \varepsilon_{C31n} \\ \varepsilon_{C321} \\ \varepsilon_{C322} \\ \cdot \\ \cdot \\ \cdot \\ \varepsilon_{C32n}
\end{bmatrix}
\qquad (3.2.24)
$$

in which $\underline{1}_c$ is written out in the first column of the matrix, $\underline{Z}_{c2}$ in the second, etc.
To obtain least-squares estimators for the β_c's the same calculation method as in the cases A and B will have to be followed. The following expressions in terms of class means are obtained; the mean calculated over the n observations of variable Y_c from the $(i,j)^{th}$ class is indicated by $\bar{Y}_{cij\cdot}$.

$$\begin{bmatrix} \hat{\beta}_{c1} \\ \hat{\beta}_{c2} \\ \hat{\beta}_{c3} \\ \hat{\beta}_{c4} \\ \hat{\beta}_{c5} \\ \hat{\beta}_{c6} \end{bmatrix} = \begin{bmatrix} \bar{Y}_{c32\cdot} \\ (\bar{Y}_{c12\cdot} - \bar{Y}_{c32\cdot}) \\ (\bar{Y}_{c22\cdot} - \bar{Y}_{c32\cdot}) \\ (\bar{Y}_{c31\cdot} - \bar{Y}_{c32\cdot}) \\ (\bar{Y}_{c11\cdot} - \bar{Y}_{c12\cdot}) - (\bar{Y}_{c31\cdot} - \bar{Y}_{c32\cdot}) \\ (\bar{Y}_{c21\cdot} - \bar{Y}_{c22\cdot}) - (\bar{Y}_{c31\cdot} - \bar{Y}_{c32\cdot}) \end{bmatrix} \qquad (3.2.25)$$

Interaction between the magnitudes 'level of education' and 'sex' is absent in the population, if within each of the classes 'lower level of education', 'medium level of education' and 'higher level of education' the mean of all Y_c-observations calculated for men and women respectively show the same difference. In that case the two classifications have, in fact, no influence on each other. In symbols this fact can be expressed by the equalities given below[21]:

$$(\bar{Y}_{c11\cdot}-\bar{Y}_{c12\cdot}) = (\bar{Y}_{c21\cdot}-\bar{Y}_{c22\cdot}) = (\bar{Y}_{c31\cdot}-\bar{Y}_{c32\cdot})$$

or

---

21. What follows again refers to the population.

$$(\bar{Y}_{C11.}-\bar{Y}_{C12.}) - (\bar{Y}_{C31.}-\bar{Y}_{C32.}) =$$

$$= (\bar{Y}_{C21.}-\bar{Y}_{C22.}) - (\bar{Y}_{C31.}-\bar{Y}_{C32.}) = 0 \qquad (3.2.26)$$

The lower line of (3.2.26) reads in terms of the parameters from (3.2.23):

$$\beta_{C5} = \beta_{C6} = 0 \qquad (3.2.27)$$

Since β_{C5} and β_{C6} are the parameters of the dummy variables $\underline{Z}_{C5}$ and $\underline{Z}_{C6}$ representing the interaction between the two classifications, it follows from (3.2.27) that in the absence of interaction the variables $\underline{Z}_{C5}$ and $\underline{Z}_{C6}$ will disappear from linear relation (3.2.23). Relation (3.2.23) then corresponds with that of case B in (3.2.15), apart from the index.

The test from regression analysis to be compared with the test from (3.1.6) will in view of the above be the significance test on β_{C5} and β_{C6}. The numerator of the F-distributed test statistic consists, in the above example, of the difference between the sum of squares of the estimated disturbances from regression (3.2.23) under the hypothesis $\beta_{C5} = \beta_{C6} = 0$ and the sum of squares of the estimated disturbances from regression (3.2.23). The denominator of the test statistic equals the latter sum of squares. This test statistic has two degrees of freedom in the numerator (equal to the number of parameters being tested), and in the denominator 6n-6 degrees of freedom (the number of observations less the number of parameters to be estimated). The statistic reads as follows, the estimated disturbances under the null hypothesis $\beta_{C5} = \beta_{C6} = 0$ being represented by the index o:

$$F_{[2,(6n-6)]} =$$

$$= \frac{(6n-6)}{2} \frac{\sum_{i=1}^{3} \sum_{j=1}^{2} \sum_{k=1}^{n} (\hat{\varepsilon}_{oijk})^2 - \sum_{i=1}^{3} \sum_{j=1}^{2} \sum_{k=1}^{n} (\hat{\varepsilon}_{cijk})^2}{\sum_{i=1}^{3} \sum_{j=1}^{2} \sum_{k=1}^{n} (\hat{\varepsilon}_{cijk})^2} \qquad (3.2.28)$$

The estimated disturbances $\hat{\varepsilon}_{oijk}$ pertain to the regression of $\underline{Y}_c$ on $\underline{1}_c$, $\underline{Z}_{c2}$, $\underline{Z}_{c3}$ and $\underline{Z}_{c4}$. This regression equals the one of (3.2.15) if index c is replaced with b. So one can obtain $\hat{\varepsilon}_{oijk}$ by substituting index b by c in (3.2.19)

$$\hat{\varepsilon}_{oijk} = Y_{cijk} - \bar{Y}_{ci..} - \bar{Y}_{c.j.} + \bar{Y}_{c...} \qquad (3.2.29)$$

Substitution of the estimated β_c's from (3.2.25) into (3.2.23) leads to the following estimated disturbances for the relation of (3.2.23):

$$\hat{\varepsilon}_{cijk} = Y_{cijk} - \bar{Y}_{cij.} \qquad (3.2.30)$$

Next substitution of the expressions for $\hat{\varepsilon}_{oijk}$ and $\hat{\varepsilon}_{cijk}$ into test statistic (3.2.28) results in:

$$F_{[2,(6n-6)]} = \frac{(6n-6)}{2} \frac{\sum_{i=1}^{3} \sum_{j=1}^{2} \sum_{k=1}^{n} (Y_{cijk} - \bar{Y}_{ci..} - \bar{Y}_{c.j.} + \bar{Y}_{c...})^2 - \sum_{i=1}^{3} \sum_{j=1}^{2} \sum_{k=1}^{n} (Y_{cijk} - \bar{Y}_{cij.})^2}{\sum_{i=1}^{3} \sum_{j=1}^{2} \sum_{k=1}^{n} (Y_{cijk} - \bar{Y}_{cij.})^2} \qquad (3.2.31)$$

The two terms in the numerator can be combined, after which (3.2.31) can ultimately be written as:

$$F_{[2,(6n-6)]} =$$

$$= \frac{(6n-6)}{2} \; \frac{n \sum_{i=1}^{3} \sum_{j=1}^{2} (\bar{Y}_{cij.} - \bar{Y}_{ci..} - \bar{Y}_{c.j.} + \bar{Y}_{c...})^2}{\sum_{i=1}^{3} \sum_{j=1}^{2} \sum_{k=1}^{n} (Y_{cijk} - \bar{Y}_{cij.})^2} \qquad (3.2.32)$$

Test statistic (3.2.32) for the test on the significance of β_{c5} and β_{c6} is - apart from the index c - equal to the test statistic on interaction from the analysis of variance (3.1.6) on page 95, the values 3 and 2 being filled in for r and s.

Naturally, other cases than the three examples elaborated here can be thought of to demonstrate the correspondence between significance tests from analysis of variance and regression analysis. The approach followed in these three examples will be directly applicable to some of those cases, or capable of being generalized. To demonstrate the connection between regression analysis and analysis of variance the examples offered here may be considered sufficient.

It is important to recognize the close relation between the techniques from regression analysis with dummy variables and analysis of variance. On the strenght of their schooling, most economists are familiar with the most important possibilities offered by regression analysis to study the relation between real variables, but as a rule they are less familiar with analysis of variance.

Analysis of variance, which is concerned first and foremost with the study of the relevance of classifications, does not, in this connection, refer to real variables but to qualitative magnitudes. Representation of qualitative magnitudes with the help of dummy variables opens the possibility of linking up regression analysis with analysis of variance. If the connection between the two methods were not recognized, then the extensive area of analysis of variance might remain closed to the economist. Communication between economists and scientists from other disciplines, which study more classification problems and hence are likely to apply analysis of variance, can be made much easier if the method of analysis commonly applied there can be translated into situations that are well-known from regression analysis.

It is possible, then, to look upon analysis of variance from the point of view of regression analysis with dummy variables, but the reverse relation, too, has been the object of study. Kendall and Stuart[22] treat regression analysis by means of tests from analysis of variance. For the most complete integration of the two methods, both the former and the latter approach would merit the attention of economists.

3.3. A COMPARISON BETWEEN ANALYSIS OF COVARIANCE AND REGRESSION ANALYSIS WITH REAL AND DUMMY VARIABLES AS REGRESSORS

In this section, proof will be given of the equivalence of, on the one hand, the significance test from analysis of covariance with one co-variable for testing a simple classification with r classes, and, on the other hand, a significance test from regression analysis. Because the relevant test statistic, see (3.1.7) on p. 97, has been derived for equal numbers of observations in each cell, namely n, the corresponsing regression must contain n elements per cell too. Analysis of covariance with unequal numbers of observations per cell are left out of consideration, but can mutatis mutandis be treated in the same

22. Ibid. p. 8, chapter 35.

way as explained below. To keep the exposition as simple as possible, only a situation with one co-variable is considered.

Before test statistic (3.1.7) can be compared to a test on the significance of a number of regression coefficients, a linear relation must be drawn up, the parameters of which are to be tested. The regressors of this relation consist of one real variable X_d[23], in this case the co-variable (see p. 96), the constant-term variable, and (r-1) binary dummy variables to represent the classification of Y into r classes. If the vectors belonging to the abovementioned variables, each containing nr observations, are denoted by, respectively, $\underline{X}_d$, $\underline{1}_d$, $\underline{Z}_{d2}$, $\underline{Z}_{d3}, \ldots, \underline{Z}_{dr}$ for the regressors, $\underline{Y}_d$ for the regressand and $\underline{\varepsilon}_d$ for the disturbances, then the linear relation with as parameters β_*, β_{d1}, $\beta_{d2}, \ldots, \beta_{dr}$, respectively, can be written as follows:

$$\underline{Y}_d = \underline{X}_d\beta_* + \underline{1}_d\beta_{d1} + \underline{Z}_{d2}\beta_{d2} + \underline{Z}_{d3}\beta_{d3} + \ldots + \underline{Z}_{dr}\beta_{dr} + \underline{\varepsilon}_d \tag{3.3.1}$$

If the kth observation (k = 1,2,...,n) from the ith class (i = 1,2,...,r) for variable Y_d is denoted by Y_{dik}, for X_d by X_{dik}, and for the disturbance by ε_{dik}, (3.3.1) can be written in full as follows:

23. The index d is only used to indicate the difference between symbols of this section and those of other sections.

$$
\begin{bmatrix}
Y_{d11} \\ Y_{d12} \\ \cdot \\ \cdot \\ \cdot \\ Y_{d1n} \\ Y_{d21} \\ Y_{d22} \\ \cdot \\ \cdot \\ \cdot \\ Y_{d2n} \\ \cdot \\ \cdot \\ \cdot \\ \cdot \\ \cdot \\ \cdot \\ \cdot \\ \cdot \\ \cdot \\ \cdot \\ \cdot \\ Y_{dr1} \\ Y_{dr2} \\ \cdot \\ \cdot \\ \cdot \\ Y_{drn}
\end{bmatrix}
=
\begin{bmatrix}
X_{d11} & 1 & 1 & 0 & & 0 \\
X_{d12} & 1 & 1 & 0 & & 0 \\
\cdot & \cdot & \cdot & \cdot & & \cdot \\
\cdot & \cdot & \cdot & \cdot & & \cdot \\
\cdot & \cdot & \cdot & \cdot & & \cdot \\
X_{d1n} & \cdot & 1 & 0 & & \cdot \\
X_{d21} & \cdot & 0 & 1 & & \cdot \\
X_{d22} & \cdot & 0 & 1 & & \cdot \\
\cdot & \cdot & \cdot & \cdot & & \cdot \\
\cdot & \cdot & \cdot & \cdot & & \cdot \\
\cdot & \cdot & \cdot & \cdot & & \cdot \\
X_{d2n} & \cdot & \cdot & 1 & & \cdot \\
\cdot & \cdot & \cdot & 0 & & \cdot \\
\cdot & \cdot & \cdot & 0 & & \cdot \\
\cdot & \cdot & \cdot & \cdot & \cdots\cdot & \cdot \\
\cdot & \cdot & \cdot & \cdot & & \cdot \\
\cdot & \cdot & \cdot & \cdot & & 0 \\
\cdot & \cdot & \cdot & \cdot & & 1 \\
\cdot & \cdot & \cdot & \cdot & & 1 \\
\cdot & \cdot & \cdot & \cdot & & \cdot \\
\cdot & \cdot & \cdot & \cdot & & \cdot \\
\cdot & \cdot & \cdot & \cdot & & \cdot \\
\cdot & \cdot & \cdot & \cdot & & 1 \\
X_{dr1} & \cdot & \cdot & \cdot & & 0 \\
X_{dr2} & \cdot & \cdot & \cdot & & 0 \\
\cdot & \cdot & \cdot & \cdot & & \cdot \\
\cdot & \cdot & \cdot & \cdot & & \cdot \\
\cdot & \cdot & \cdot & \cdot & & \cdot \\
X_{drn} & 1 & 0 & 0 & & 0
\end{bmatrix}
\begin{bmatrix}
\beta_{*} \\ \beta_{d1} \\ \beta_{d2} \\ \cdot \\ \cdot \\ \cdot \\ \beta_{dr}
\end{bmatrix}
+
\begin{bmatrix}
\varepsilon_{d11} \\ \varepsilon_{d12} \\ \cdot \\ \cdot \\ \cdot \\ \varepsilon_{d1n} \\ \varepsilon_{d21} \\ \varepsilon_{d22} \\ \cdot \\ \cdot \\ \cdot \\ \varepsilon_{d2n} \\ \cdot \\ \cdot \\ \cdot \\ \cdot \\ \cdot \\ \cdot \\ \cdot \\ \cdot \\ \cdot \\ \cdot \\ \cdot \\ \varepsilon_{dr1} \\ \varepsilon_{dr2} \\ \cdot \\ \cdot \\ \cdot \\ \varepsilon_{drn}
\end{bmatrix}
\qquad (3.3.2)
$$

In the second column of matrix (3.3.2) $\underline{1}_d$ is written out, in the third column $\underline{Z}_{d2}$, etc. If the matrix with dummy variables is called $\underline{W}_d$ and the vector with corresponding parameters $\underline{\beta}_d$, then (3.3.1) can be written in a much more compact form as

$$\underline{Y}_d = \underline{X}_d \beta_* + \underline{W}_d \underline{\beta}_d + \underline{\varepsilon}_d \qquad (3.3.3)$$

Before the null hypothesis of a non-significant classification can be transformed into a null hypothesis relevant to the parameters of (3.3.1), the meaning of these parameters must be known. To determine the regression coefficients the results on pp. 35-38 can be put to use with success. It was demonstrated there that the coefficients of the regression of $\underline{Y}$ on real variables, grouped into matrix $\underline{X}$, and dummy variables, denoted by matrix $\underline{W}$, can be obtained with the help of the estimated disturbances from the regressions of $\underline{Y}$ on $\underline{W}$ and of $\underline{X}$ on $\underline{W}$. For the linear relation (3.3.1) this means that the estimated disturbances of the regression of $\underline{Y}_d$ on $\underline{W}_d$ and of $\underline{X}_d$ on $\underline{W}_d$ must be determined.
A regression analogous to that of $\underline{Y}_d$ and that of $\underline{X}_d$ on the dummy variables from (3.3.1) is to be found in (3.2.1) on page 99. The relevant estimated disturbances are mentioned in (3.2.12) on page 104. On the analogy of (3.2.12) the following result[24] is reached for the estimated disturbances from the two regressions (analogously to the method followed on pp. 35-38, the estimated disturbance is given an index c).

$\hat{Y}_{cdik} = \bar{Y}_{dik} - \bar{Y}_{di.}$, the $(i,k)^{th}$ estimated disturbance from the regression of $\underline{Y}_d$ on $\underline{W}_d$

24. The mean calculated over the observations of the ith class is represented for the variables Y_d and X_d by $\bar{Y}_{di.}$ and $\bar{X}_{di.}$ respectively.

$\hat{X}_{cdik} = X_{dik} - \bar{X}_{di.}$, the $(i,k)^{th}$ estimated disturbance from the regression of $\underline{X}_d$ on $\underline{W}_d$.

(3.3.4)

As can be derived from (2.2.19) on page 37, the estimator for paramater β_*, which is comparable to $\underline{\alpha}$ from (2.2.12) on page 35, is obtained by means of simple regression of $\hat{Y}_{cdik}$ on $\hat{X}_{cdik}$. Expressed in the symbols of (3.3.4) the estimator of parameter β_* will then read:

$$\hat{\beta}_* = \frac{\sum_{i=1}^{r} \sum_{k=1}^{n} (Y_{dik} - \bar{Y}_{di.})(X_{dik} - \bar{X}_{di.})}{\sum_{i=1}^{r} \sum_{k=1}^{n} (X_{dik} - \bar{X}_{di.})^2} \tag{3.3.5}$$

The estimators for the parameters $\beta_{d1}, \beta_{d2}, \ldots, \beta_{dr}$ from (3.3.1) are found quite simply with the help of (2.2.21) on page 37; they read:

$$\begin{aligned}
\hat{\beta}_{d1} &= \bar{Y}_{dr.} - \bar{X}_{dr.}\hat{\beta}_* \\
\hat{\beta}_{d2} &= (\bar{Y}_{d1.} - \bar{Y}_{dr.}) - (\bar{X}_{d1.} - \bar{X}_{dr.})\hat{\beta}_* \\
\hat{\beta}_{d3} &= (\bar{Y}_{d2.} - \bar{Y}_{dr.}) - (\bar{X}_{d2.} - \bar{X}_{dr.})\hat{\beta}_* \\
&\;\vdots \\
\hat{\beta}_{dr} &= (\bar{Y}_{d(r-1).} - \bar{Y}_{dr.}) - (\bar{X}_{d(r-1).} - \bar{X}_{dr.})\hat{\beta}_*
\end{aligned} \tag{3.3.6}$$

Before testing the classification on significance, a correction is applied in analysis of covariance for the influence of co-variable X_d on the variable Y_d to be examined. W.G. Cochran[25] uses as corrected values of Y_d:

---

25. Ibid. p. 96, p. 269.

(the corrected value is indicated by an asterisk on the upper right-hand side):

$$Y^{*}_{dik} = Y_{dik} - X_{dik}\hat{\beta}_{*} \tag{3.3.7}$$

For the above corrected variable the significance of the classification into r classes is tested in (3.1.7) on page 97. Using (3.3.7), one can relate the null hypothesis in respect of the parameters from (3.3.6) in the following manner:

If for (3.3.7) the mean for the ith category is calculated, one obtains:

$$\bar{Y}^{*}_{di.} = \bar{Y}_{di.} - \bar{X}_{di.}\hat{\beta}_{*} \tag{3.3.8}$$

Combination of (3.3.8) and (3.3.6) leads to the following expressions for the estimated parameters of (3.3.1):

$$\begin{aligned}
\hat{\beta}^{*}_{d1} &= \bar{Y}^{*}_{dr.} \\
\hat{\beta}^{*}_{d2} &= \bar{Y}^{*}_{d1.} - \bar{Y}^{*}_{dr.} \\
\hat{\beta}^{*}_{d3} &= \bar{Y}^{*}_{d2.} - \bar{Y}^{*}_{dr.} \\
&\ \ \vdots \\
\hat{\beta}^{*}_{dr} &= \bar{Y}^{*}_{d(r-1)} - \bar{Y}^{*}_{dr.}
\end{aligned} \tag{3.3.9}$$

The null hypothesis of the significance test on the classification of the corrected Y_d-variable reads[26]:

---

26. What follows refers of the population.

$$\bar{Y}^*_{d1.} = \bar{Y}^*_{d2.} = \ldots = \bar{Y}^*_{dr.}$$

or (3.3.10)

$$\bar{Y}^*_{d1.} - \bar{Y}^*_{dr.} = \bar{Y}^*_{d2.} - \bar{Y}^*_{dr.} = \ldots = \bar{Y}^*_{d(r-1).} - \bar{Y}^*_{dr.} = 0$$

For the parameters from (3.3.1), null hypothesis (3.3.10) implies that given (3.3.9), the following null hypothesis should be stated:

$$\beta_{d2} = \beta_{d3} = \ldots = \beta_{dr} = 0 \tag{3.3.11}$$

We will now examine whether the usual test on the significance of the parameters form (3.3.11) yields a test statistic that is equivalent with the one in (3.1.7). This familiar test on the simultaneous significance of a number of parameters has been used, among other places, in section 3.2., see (3.2.28) on page 116. For this test one needs the estimated disturbances from relation (3.3.1) as well as those under the null hypothesis (3.3.11). These estimated disturbances must be calculated for the linear relation given below[27]:

$$\underline{Y}_d = \underline{X}_d\underline{\beta}_{*0} + \underline{1}_d\beta_{d10} + \underline{\varepsilon}_{d0} \tag{3.3.12}$$

The estimated disturbances for relation (3.3.1) are obtained by substituting (3.3.6) into (3.3.1) for $\beta_{d1}, \beta_{d2}, \ldots, \beta_{dr}$, and substituting $\hat{\beta}_*$ for β_*. The $(i,k)^{th}$ estimated disturbance reads:

$$\hat{\varepsilon}_{dik} = (Y_{dik} - \bar{Y}_{di.}) - (X_{dik} - \bar{X}_{di.})\hat{\beta}_* \tag{3.3.13}$$

27. The subscript 0 indicates that (3.3.1) is considered under the null hypothesis (3.3.11).

The estimated disturbances for (3.3.12) are obtained as follows, after the simple regression of $\underline{Y}_d$ on $\underline{X}_d$ has produced the estimators $\hat{\beta}_{*0}$ and $\hat{\beta}_{d10}$:

$$\hat{\varepsilon}_{dik0} = Y_{dik} - X_{dik}\beta_{*0} - \beta_{d10} = Y_{dik} - \bar{Y}_{d..} - (X_{dik} - \bar{X}_{d..})\beta_{*0} \tag{3.3.14}$$

because the estimator for the constant β_{d10} is, in simple regression:

$$\hat{\beta}_{d10} = \bar{Y}_{d..} - \bar{X}_{d..}\hat{\beta}_{*0} \tag{3.3.15}$$

Substituting $\hat{\varepsilon}_{0ijk}$ in (3.2.28) by expression (3.3.14) for $\hat{\varepsilon}_{dik0}$, and $\hat{\varepsilon}_{cijk}$ by formula (3.3.13), and adjusting the number of degrees of freedom to the situation in the present subsection, one obtains the test statistic:

$$F_{[(r-1), rn-(r+1)]} = \frac{[rn-(r+1]}{(r-1)} \quad \frac{A-B-C+D}{C-D} \tag{3.3.16}$$

where

$$A = \sum_{i=1}^{r} \sum_{k=1}^{n} (Y_{dik} - \bar{Y}_{d..})^2, \quad B = (\hat{\beta}_{*0})^2 \sum_{i=1}^{r} \sum_{k=1}^{n} (X_{dik} - \bar{X}_{d..})^2$$

$$C = \sum_{i=1}^{r} \sum_{k=1}^{n} (Y_{dik} - \bar{Y}_{di.})^2 \text{ and}$$

$$D = (\hat{\beta}_{*})^2 \sum_{i=1}^{r} \sum_{k=1}^{n} (X_{dik} - \bar{X}_{di.})^2$$

Apart from index d, which does not mean a real difference, the test statistics from (3.3.16) and (3.1.7) are identical, for according to (3.1.1) see page 92, it is true that:

$$\sum_{i=1}^{r} \sum_{k=1}^{n} (Y_{dik} - \bar{Y}_{d..})^2 - \sum_{i=1}^{r} \sum_{k=1}^{n} (Y_{dik} - \bar{Y}_{di.})^2 =$$

$$= n \sum_{i=1}^{r} (\bar{Y}_{di.} - \bar{Y}_{d..})^2 \qquad (3.3.17)$$

It has now been proved that the test from the analysis of covariance on the significance of a classification corresponds with the test on the significance of the parameters of the binary dummy variables from the relevant regression. This correspondence between the two methods has a number of advantages. First, relating analysis of covariance and regression analysis is illuminating, because the two methods are in most cases taught and applied quite unconnectedly. Analysis of covariance is applied in those disciplines where experimenting is the rule; the use of regression analysis in economics is natural, because of the lack of opportunities to experiment. Communication between scientists of different disciplines is promoted when they understand one another's research methods and moreover look for any relations between them. Second, the analogy found to exist between regression analysis and analysis of covariance has the practical advantage that analysis of covariance can be carried out with the equipment available for regression analysis[28]. Last, the relation between the two methods opens the way to fast transmission of new theoretical developments from one method to the other.

---

28. In this context we may think of cumputer programmes available for the computation of regressions.

4. Binary dummy variables as a help in examining the constancy of parameters in linear relations

4.1. INTRODUCTION; CHOW'S TEST

Estimating parameters from relations between economic variables is in general performed under the assumption that the relevant parameters are constant for all observations. The hypothesis of constancy of parameters, mostly not made explicit, will in most cases be acceptable for studies based on short time series. If the time series extend over several decades, then the a priori assumption of the constancy of the parameters becomes disputable, for owing to structural shifts in economic life the parameters can gradually change their values. Gradual changes in parameters can only be quantified in a model if the analytical form of their movement is known. Since in most cases it is not, another method will have to be followed to quantify changes in parameters. One may, for instance, try to subdivide the whole observation period into periods that are homogeneous in respect of the value of the parameters, and then study the path of the parameter estimates from period to period. Subdivision into homogeneous subperiods are in most cases subjective, however[1]. Moreover it appears that parameter estimates are very sensitive to the manner of subdividing the observations. If there is not a more accurate indication about the nature of the movement in the parameters, research threatens to come to a dead end.

1. See G. Menges and H. Diehl, Time stability of structural parameters, Colston Papers no. 16, pp. 299-320.

There are cases, however, where a specific subdivision of the observation material follows naturally from economic considerations. Such a case occurs, for example, when data are available about a pre-war and a post-war period or if there are observations from war years and peace years. In that case it is possible to retain that subdivision for examining the constancy of the parameters. With quarterly data the obvious course is to investigate whether or not the same set of parameters may be used for each of the individual quarters; the total period may then be distinguished as four subperiods. The contents of the coming sections will be limited to those cases in which clear motives can be given for such a subdivision of the observation material. The hypothesis of constancy of the parameters will then be tested by verifying that the parameters of each subperiod do not deviate significantly from each other.
One of the tests on the equality of parameters from linear relations for two separate periods is known under the name of Chow's test[2]. Chow has examined whether or not for both periods the same set of parameter estimates may be used. If the test does not point towards discarding the null hypothesis of equal parameters, then one may proceed to the estimation of the parameters on the ground of all observations taken together. If it turns out, however, that the null hypothesis is to be rejected, then one will find oneself compelled to continue with regressions for the individual periods, unless there are considerations of economic theory that warrant the introduction of entirely different models. A drawback to this method is the, often heavy, loss of degrees of freedom suffered in estimating and testing. If one wants,

---

2. Ibid. p. 11.

for example, to make a forecast for 1976, and the test has made it clear that pre-war observations have a different parameter structure from the post-war observations, then one can use only the post-war data for predicting the dependent variable in 1976, thus losing in the process a number of degrees of freedom equalling the number of pre-war observations.

The loss of degrees of freedom can be limited by a different approach. With the help of dummy variables[3] - how, will be explained in the next section - it becomes possible to change each parameter at the transition from one period to another. The gain in degrees of freedom can be quite considerable, because now all observations are put to use.

The procedure of introducing discontinuous shifts in parameters by means of binary dummy variables has been known a long time. From the literature[4] we know about the consumption function, which has different parameters in periods of peace and of war. The linear relations read:

$$
\begin{aligned}
C_t &= \alpha_1 + \beta_1 Y_t + \varepsilon_t \quad \text{in years of war} \\
C_t &= \alpha_2 + \beta_2 Y_t + \varepsilon_t \quad \text{in years of peace}
\end{aligned}
\qquad (4.1.1)
$$

Next, the two relations from (4.1.1), are combined with the help of the binary dummy variable Z, which assumes the value zero for war years and the value one for peace years:

$$C_t = \alpha_1 + \gamma_1 Z_t + \beta_1 Y_t + \gamma_2 Z_t Y_t + \varepsilon_t \qquad (4.1.2)$$

3. See the comment on the article of Menges and Diehl, ibid. p. 128, p. 318.
4. See J. Johnston, ibid. p. 8, pp. 177 ff.

The symbols γ_1 and γ_2 have the following meaning:

$$\gamma_1 = \alpha_2 - \alpha_1 \quad \text{and} \quad \gamma_2 = \beta_2 - \beta_1 \tag{4.1.3}$$

The relation between the parameters from (4.1.2) and (4.1.1), as written down in (4.1.3) is also true of the least-squares estimators of the relevant parameters; in fact, in section 4.2 the following equalities will be proved:

$$\hat{\gamma}_1 = \hat{\alpha}_2 - \hat{\alpha}_1 \quad \text{and} \quad \hat{\gamma}_2 = \hat{\beta}_2 - \hat{\beta}_1 \tag{4.1.3a}$$

If parameter γ_1 is tested for significance, a significance test is in fact carried out on $\alpha_2 - \alpha_1$. In the same way, testing the significance of γ_2 implies a significance test on the difference between β_2 and β_1 [5]. In section 4.2 it will be proved that the simultaneous test on the significance of γ_1 and γ_2 is equal to the test of Chow.

An important advantage of making use of dummy variables in the manner described above is the possibility of indicating for each individual parameter whether there is a significant difference between the values it assumes in the two periods. Chow's test on the contrary does not go beyond a statement about all the parameters simultaneously; it is not possible by that method to indicate which parameters show significant differences. Yet for the economist who is investigating possible structural changes it will be interesting to know what are the variables whose parameters change significantly at the

5. See also Damodar Gurajati, Use of dummy variables in testing for equality between sets of coefficients in two linear regressions: a note, The American Statistican, Vol. 24 (1970), no. 1, pp. 50-52.

transition from one period to another.
After the parameters that vary significantly from one period to another have been identified, it often suffices, in drawing up the regression equation in its ultimate form just to include a dummy variable for these parameters. The number of degrees of freedom ultimately lost as compared with a regression based on the (faulty) assumption of constant parameters is thus reduced to less than the number of parameters.
Now a more detailed explanation of Chow's test will be presented. This test examines the simultaneous equality of corresponding parameters for two subperiods. If these subperiods are denoted by the indices 1 and 2, the linear relations between the first T_1 observations and the last T_2 observations ($T_1+T_2 = T$), respectively, can be written as

$$\begin{aligned} \underline{Y}_1 &= \underline{X}_1\underline{\beta}_1 + \underline{\varepsilon}_1 \\ \underline{Y}_2 &= \underline{X}_2\underline{\beta}_2 + \underline{\varepsilon}_2 \end{aligned} \tag{4.1.4}$$

or

$$\begin{bmatrix} \underline{Y}_1 \\ \underline{Y}_2 \end{bmatrix} = \begin{bmatrix} \underline{X}_1 & \underline{0} \\ \underline{0} & \underline{X}_2 \end{bmatrix} \begin{bmatrix} \underline{\beta}_1 \\ \underline{\beta}_2 \end{bmatrix} + \begin{bmatrix} \underline{\varepsilon}_1 \\ \underline{\varepsilon}_2 \end{bmatrix} \tag{4.1.4a}$$

where:

$\underline{Y}_1$: vector of the first T_1 observations of the regressand

$\underline{X}_1$: matrix of the first T_1 observations of the Λ regressors

$\underline{\beta}_1$: vector of the parameters from the first period

$\underline{\varepsilon}_1$: vector of the first T_1 disturbances.

The same holds, mutatis mutandis, for index 2.
Both T_1 and T_2 must be larger than Λ - the number of parameters including the constant - in order to allow for an estimation of the parameters in each of the subperiods. If this condition is not fulfilled, the possibility of directly examining the equality of $\underline{\beta}_1$ and $\underline{\beta}_2$ is ruled out, because one of the two cannot be estimated. Moreover in this chapter the assumptions 1, 2 and 4 of pages 18 and 19 are made with respect to the disturbances.
The null hypothesis made by Chow reads in the symbols of (4.1.4a):

$$\underline{\beta}_1 = \underline{\beta}_2 = \underline{\beta}_0 \tag{4.1.5}$$

Under the null hypothesis relation (4.1.4a) passes into[6]

$$\begin{bmatrix} \underline{Y}_1 \\ \underline{Y}_2 \end{bmatrix} = \begin{bmatrix} \underline{X}_1 \\ \underline{X}_2 \end{bmatrix} \underline{\beta}_0 + \begin{bmatrix} \underline{\varepsilon}_{10} \\ \underline{\varepsilon}_{20} \end{bmatrix} \tag{4.1.6}$$

In constructing his test Chow then starts from the identity

$$\begin{bmatrix} \underline{Y}_1 - \underline{X}_1\hat{\underline{\beta}}_0 \\ \underline{Y}_2 - \underline{X}_2\hat{\underline{\beta}}_0 \end{bmatrix} = \begin{bmatrix} \underline{Y}_1 - \underline{X}_1\hat{\underline{\beta}}_1 \\ \underline{Y}_2 - \underline{X}_2\hat{\underline{\beta}}_2 \end{bmatrix} + \begin{bmatrix} \underline{X}_1(\hat{\underline{\beta}}_1 - \hat{\underline{\beta}}_0) \\ \underline{X}_2(\hat{\underline{\beta}}_2 - \hat{\underline{\beta}}_0) \end{bmatrix} \tag{4.1.7}$$

Left of the equality sign in (4.1.7) stands the vector of estimated disturbances of the relation (4.1.6). The first term to the right of the equality sign contains the estimated disturbances of (4.1.4a); the second term is built up as a linear combination of the vectors consisting of observations of the regressors. A well-known

6. The index zero is added to the disturbance term so as to distinguish the disturbances in (4.1.4a) from those in (4.1.6).

result in regression analysis is[7] that the estimated disturbances of a relation are uncorrelated with the regressors from that relation, which implies that the vectors to the right of the equality sign are uncorrelated.

The sum of the squares of the estimated disturbances in relation (4.1.6) corresponds with the inner product[8] of the vector to the left of the equality sign in (4.1.7). For this inner product the following relation holds:

$$\left\| \begin{matrix} \underline{Y}_1 - \underline{X}_1 \underline{\hat{\beta}}_0 \\ \\ \underline{Y}_2 - \underline{X}_2 \underline{\hat{\beta}}_0 \end{matrix} \right\|^2 = \left\| \begin{matrix} \underline{Y}_1 - \underline{X}_1 \underline{\hat{\beta}}_1 \\ \\ \underline{Y}_2 - \underline{X}_2 \underline{\hat{\beta}}_2 \end{matrix} \right\|^2 + \left\| \begin{matrix} \underline{X}_1 (\underline{\hat{\beta}}_1 - \underline{\hat{\beta}}_0) \\ \\ \underline{X}_2 (\underline{\hat{\beta}}_2 - \underline{\hat{\beta}}_0) \end{matrix} \right\|^2 \qquad (4.1.8)$$

To the right of the equality sign in (4.1.8) only squares figure, because the relevant vectors are uncorrelated. For brevity, Chow writes expression (4.1.8) as follows:

$$Q_1 = Q_2 + Q_3 \qquad (4.1.9)$$

Chow then uses a test statistic of the form:

$$F_{[\Lambda, T-2\Lambda]} = \frac{(T-2\Lambda)}{\Lambda} \frac{Q_3}{Q_2} \qquad (4.1.10)$$

The F-distribution may be applied because Q_3 and Q_2 are independent. Since Q_2 represents the sum of squares of the estimated disturbances from (4.1.4a), this quantity has $T - 2\Lambda$ degrees of freedom. The statistic Q_3 has Λ degrees of freedom; this can be verified by examining (4.1.9).

---

7. See A.S. Goldberger, ibid. p. 7, p. 159.
8. The in-product of a vector Y is denoted by $\|Y\|^2$, which equals $\sum_{t=1}^{T} Y_t^2$.

The numerator in (4.1.10) will be close to zero under the null hypothesis. The value of Q_3 increases ceteris paribus for an increasing difference between β_1 and β_0 and between β_2 and β_0.
Chow remarks with respect to the test statistic in (4.1.10) that it corresponds with the test statistic from analysis of covariance, applied in the manner described in section 3.1 for testing the equality of parameters in linear relations. At the end of section 4.2 this remark will be dealt with in more detail by means of an example given by Malinvaud.
In his article Chow also recognizes the desirability of being able to test part of the parameters for equality in different periods. From other investigations it is sometimes known a priori that one or some of the parameters do not remain constant for the whole period under observation (the relevant parameters and regressors will be provided with an index a); only the remaining ones (to be denoted by an index b) must be tested on constancy. Chow writes the linear relations as follows:

$$\begin{aligned} \underline{Y}_1 &= \underline{X}_{1a}\underline{\beta}_{1a} + \underline{X}_{1b}\underline{\beta}_{1b} + \underline{\varepsilon}_1 \\ \underline{Y}_2 &= \underline{X}_{2a}\underline{\beta}_{2a} + \underline{X}_{2b}\underline{\beta}_{2b} + \underline{\varepsilon}_2 \end{aligned} \tag{4.1.11}$$

and

$$\begin{bmatrix} \underline{Y}_1 \\ \underline{Y}_2 \end{bmatrix} = \begin{bmatrix} \underline{X}_{1a} & \underline{0} & \underline{X}_{1b} & \underline{0} \\ \underline{0} & \underline{X}_{2a} & \underline{0} & \underline{X}_{2b} \end{bmatrix} \begin{bmatrix} \underline{\beta}_{1a} \\ \underline{\beta}_{2a} \\ \underline{\beta}_{1b} \\ \underline{\beta}_{2b} \end{bmatrix} + \begin{bmatrix} \underline{\varepsilon}_1 \\ \underline{\varepsilon}_2 \end{bmatrix} \tag{4.1.11a}$$

where:

$\underline{Y}_1$: vector of the first T_1 observations of the regressand

$\underline{X}_{1a}$: matrix of the first T_1 observations of the regressors which correspond with non-constant parameters

$\underline{X}_{1b}$: matrix of the first T_1 observations of the regressors which correspond with parameters the constancy of which is to be tested

$\underline{\beta}_{1a}$: vector of parameters from the first periods which are known to differ from the corresponding parameters from the second period $\underline{\beta}_{2a}$

$\underline{\beta}_{1b}$: vector of parameters from the first period of which one wishes to check their equality with corresponding parameters from the second period $\underline{\beta}_{2b}$

$\underline{\varepsilon}_1$: vector of the first T_1 disturbances.

The same for index 2. The number of parameters in $\underline{\beta}_{1a}$ and $\underline{\beta}_{1b}$ is indicated by $(\Lambda-p)$ and p, respectively. Consequently, $\underline{X}_{1a}$ is of the order $T_1 \times (\Lambda-p)$ and $\underline{X}_{1b}$ of the order $T_1 \times p$.

The null hypothesis for the situation sketched above looks as follows:

$$\underline{\beta}_{1b} = \underline{\beta}_{2b} = \underline{\beta}_{b0} \qquad (4.1.12)$$

Under the null hypothesis, (4.1.11a) passes into[9];

9. To make a distinction between the situation in (4.1.11a) and that under the null hypothesis, the disturbance vectors and the non-constant parameters are given an index 0.

$$\begin{bmatrix} \underline{Y}_1 \\ \underline{Y}_2 \end{bmatrix} = \begin{bmatrix} \underline{X}_{1a} & \underline{0} & \underline{X}_{1b} \\ \underline{0} & \underline{X}_{2a} & \underline{X}_{2b} \end{bmatrix} \begin{bmatrix} \underline{\beta}_{1a0} \\ \underline{\beta}_{2a0} \\ \underline{\beta}_{b0} \end{bmatrix} + \begin{bmatrix} \underline{\varepsilon}_{10} \\ \underline{\varepsilon}_{20} \end{bmatrix} \qquad (4.1.13)$$

Analogously with the previous case (see 4.1.7), in which the (Λ-p) non-constant parameters did not occur, Chow starts from the equality expressed below:

$$\begin{bmatrix} \underline{Y}_1 - \underline{X}_{1a}\hat{\underline{\beta}}_{1a0} - \underline{X}_{1b}\hat{\underline{\beta}}_{b0} \\ \underline{Y}_2 - \underline{X}_{2a}\hat{\underline{\beta}}_{2a0} - \underline{X}_{2b}\hat{\underline{\beta}}_{b0} \end{bmatrix} = \begin{bmatrix} \underline{Y}_{1a} - \underline{X}_{1a}\hat{\underline{\beta}}_{1a} - \underline{X}_{1b}\hat{\underline{\beta}}_{1b} \\ \underline{Y}_{2a} - \underline{X}_{2a}\hat{\underline{\beta}}_{2a} - \underline{X}_{2b}\hat{\underline{\beta}}_{2b} \end{bmatrix} + \begin{bmatrix} \underline{X}_{1a}(\hat{\underline{\beta}}_{1a} - \hat{\underline{\beta}}_{1a0}) + \underline{X}_{1b}(\hat{\underline{\beta}}_{1b} - \hat{\underline{\beta}}_{1b0}) \\ \underline{X}_{2a}(\hat{\underline{\beta}}_{2a} - \hat{\underline{\beta}}_{2a0}) + \underline{X}_{2b}(\hat{\underline{\beta}}_{2b} - \hat{\underline{\beta}}_{2b0}) \end{bmatrix}$$

(4.1.14)

As in (4.1.7), the left-hand term of (4.1.14) has the maining of estimated disturbances, viz. those of the relation (4.1.13). In the same way, the first vector to the right of the equality sign in (4.1.14) has as its elements the estimated disturbances of the relation (4.1.11a). For the same reasons as have been put forward with (4.1.7), the vectors from the right-hand term of (4.1.14) show no mutual correlation. Starting from the identity of (4.1.14) Chow arrives at the expression analogous to (4.1.9):

$$Q'_1 = Q'_2 + Q'_3 \qquad (4.1.15)$$

The sum of squares Q'_1 has T - (2Λ-p) degrees of freedom because in (4.1.13) 2Λ-p parameters have to be estimated. Quantity Q'_2 has T-2Λ degrees of freedom, because in (4.1.11) 2Λ parameters have to be estimated before one can calculate the estimated disturbances. Quantity Q'_3 consequently has a number of degrees of freedom equal to T - (2Λ-p) - (T-2Λ) = p.

Chow uses the following test statistic, built up on the analogy of (4.1.10):

$$F_{[p,T-2\Lambda]} = \frac{(T-2\Lambda)}{p} \cdot \frac{Q'_3}{Q'_2} \qquad (4.1.16)$$

In section 4.2 an alternative derivation will be given of the test statistic from (4.1.10) as well as the one from (4.1.16), in which they are obtained by means of binary dummy variables.

Chow's test, in the form presented by him, can be applied only when the observation material falls into two parts. With quarterly data, however, there is a division into four subsets, while a division into three parts can be desirable for data from a pre-war period, the war years and the post-war years. In section 4.2 a generalization of Chow's test will be presented for a split-up into four periods. As in the alternative derivation of Chow's test, binary dummy variables will be applied there. It will become evident that the test for equality of the parameters over four subperiods offers a better approach than the method followed by Klein[10], because the subjective element will be eliminated from the procedure.

Although the power of the test constitutes a very important consideration in judging its usefulness, Chow gives no indication whatsoever. The alternative derivation of Chow's test, to be described below, offers a simple way to determine the power of the test. In section 4.3, the distribution of the test statistic under an alternative hypothesis will be derived.

---

10. A desription of Klein's method is to be found on pages 74 and 75.

4.2. AN ALTERNATIVE DERIVATION OF CHOW'S TEST WITH THE HELP OF BINARY DUMMY VARIABLES

One may wonder if the coefficients obtained in one regression which uses all the observations available from two clearly distinguishable periods, can have any real meaning. If the regression coefficients for each individual subperiod show large differences, one obtains for coefficients estimated with the help of all available observations collectively a meaningless average of the coefficients from the regressions on the subperiods. In this situation the coefficients relevant to the whole period describe neither the structure in the first, nor that in the second period. Tests carried out on these coefficients and their use for analysing economic-political problems have no meaning either, because the condition of homogeneity of the observation material has not been fulfilled.

Before combining subperiods that are clearly distinguishable to the economist, one should, therefore, test the hypothesis that the parameters from the two subperiods do not differ significantly. Chow developed for that purpose the test mentioned and described already in section 4.1. But this test can also be developed in another manner, viz. with the help of binary dummy variables. The alternative derivation will be presented below. Next it will be demonstrated that this new approach offers the possibility of extending Chow's test.

The linear relations between the vector with regressand values $\underline{Y}$ and the matrix with regressors $\underline{X}$ which have to be tested for equality of parameters, read (see (4.1.4)):

$$\begin{aligned} &\text{for the first period} \quad && \underline{Y}_1 = \underline{X}_1 \underline{\beta}_1 + \underline{\varepsilon}_1 \\ &\text{for the second period} \quad && \underline{Y}_2 = \underline{X}_2 \underline{\beta}_2 + \underline{\varepsilon}_2 \end{aligned} \qquad (4.2.1)$$

The two relations in (4.2.1) can be written as one relation as follows:

$$\begin{bmatrix} \underline{Y}_1 \\ \underline{Y}_2 \end{bmatrix} = \begin{bmatrix} \underline{X}_1 & \underline{0} \\ \underline{0} & \underline{X}_2 \end{bmatrix} \begin{bmatrix} \underline{\beta}_1 \\ \underline{\beta}_2 \end{bmatrix} + \begin{bmatrix} \underline{\varepsilon}_1 \\ \underline{\varepsilon}_2 \end{bmatrix} \qquad (4.2.1a)$$

Because later on in this section the least-squares estimator for the parameter vector from (4.2.1a) will be used, we shall first comment on that estimator. By applying formula (2.1.7) see p. 19, to the relation in (4.2.1a), one obtains, after substituting for the symbols $\underline{Y}$, $\underline{X}$ and $\hat{\underline{\beta}}$ of (2.1.7), respectively, $\begin{bmatrix} \underline{Y}_1 \\ \underline{Y}_2 \end{bmatrix}$, $\begin{bmatrix} \underline{X}_1 & \underline{0} \\ \underline{0} & \underline{X}_2 \end{bmatrix}$ and $\begin{bmatrix} \hat{\underline{\beta}}_1 \\ \hat{\underline{\beta}}_2 \end{bmatrix}$, the following estimator for the parameters in (4.2.1a):

$$\begin{bmatrix} \hat{\underline{\beta}}_1 \\ \hat{\underline{\beta}}_2 \end{bmatrix} = \begin{bmatrix} (\underline{X}_1'\underline{X}_1)^{-1} & \underline{X}_1'\underline{Y}_1 \\ (\underline{X}_2'\underline{X}_2)^{-1} & \underline{X}_2'\underline{Y}_2 \end{bmatrix} \qquad (4.2.2)$$

From (4.2.2) it appears that the estimated parameters of (4.2.1a) correspond with the least-squares estimator of the parameters of, respectively, the relation between $\underline{Y}_1$ and $\underline{X}_1$ and the relation between $\underline{Y}_2$ and $\underline{X}_2$. The combination of the relations in (4.2.1) into the one relation of (4.2.1a) apparently has no consequences for the estimators of the parameters. The idea of combining two relations into one will be taken up presently, when with the help of a binary dummy variable a relation will be evolved which has $\underline{\beta}_1$ and $(\underline{\beta}_2-\underline{\beta}_1)$ as parameters.
Section 4.1 contains an exposition of the way in which a discontinuous change in a parameter can be represented by means of a binary dummy variable. In a analogous way the relations from (4.2.1) for the first and second periods, viz.

$$Y_t = \beta_{11} + X_{1t}\beta_{12} + X_{2t}\beta_{13} + \ldots + X_{(\Lambda-1)t}\beta_{1\Lambda} + \varepsilon_t$$

for $t = 1,2,\ldots,T_1$, and

$$Y_t = \beta_{21} + X_{1t}\beta_{22} + X_{2t}\beta_{23} + \ldots + X_{(\Lambda-1)t}\beta_{2\Lambda} + \varepsilon_t \tag{4.2.3}$$

for $t = T_1 + 1,\ T_1 + 2,\ldots,T$, respectively

can be taken together with the help of the binary dummy variable Z_t, which is equal to zero for $t = 1,2,\ldots,T_1$ and equal to one for $t = T_1 + 1,\ T_1 + 2,\ldots,T$:

$$Y_t = \beta_{11} + X_{1t}\beta_{12} + X_{2t}\beta_{13} + \ldots + X_{(\Lambda-1)t}\beta_{1\Lambda} + \\ + \gamma_1 Z_t + \gamma_2 Z_t X_{1t} + \gamma_3 Z_t X_{2t} + \ldots + \gamma_\Lambda Z_t X_{(\Lambda-1)t} + \varepsilon_t \tag{4.2.4}$$

where γ_i $(i = 1,2,\ldots,\Lambda)$ indicates the change in the ith parameter at the transition from the first to the second-period. It is simple to demonstrate that (4.2.4) can be written in matrix notation as:

$$\begin{bmatrix} \underline{Y}_1 \\ \underline{Y}_2 \end{bmatrix} = \begin{bmatrix} \underline{X}_1 & \underline{0} \\ \underline{X}_2 & \underline{X}_2 \end{bmatrix} \begin{bmatrix} \underline{\beta}_1 \\ \underline{\gamma} \end{bmatrix} + \begin{bmatrix} \underline{\varepsilon}_1 \\ \underline{\varepsilon}_2 \end{bmatrix} \tag{4.2.5}$$

(See for the meaning of the symbols the explanation under (4.1.4a)). The difference between (4.2.5) and (4.2.1a) consists in a change of the matrix of regressors, in which this time matrix $\underline{X}_2$ figures twice in the bottom row. As in (4.1.3), it is true that the meaning of relation (4.2.5) corresponds with that of relation (4.2.1a), if

$$\underline{\gamma} = \underline{\beta}_2 - \underline{\beta}_1 \qquad (4.2.6)$$

On the basis of (4.2.5) it will be demonstrated successively that it is true of the least-squares estimator of $\underline{\gamma}$ that:

$$\hat{\underline{\gamma}} = \hat{\underline{\beta}}_2 - \hat{\underline{\beta}}_1 \qquad (4.2.6a)$$

and that the significance test on $\underline{\gamma}$ corresponds with Chow's test.

Equality (4.2.6a) can be proved by deriving the least-squares estimator for $\underline{\beta}_1$ and $\underline{\gamma}$ from (4.2.5) in a way analogous to the one producing (4.2.2). There is a simpler method, however. Starting from the matrix of the regressors from (4.2.1a) one arrives, using a matrix $\underline{P}$ which is equal to $\begin{bmatrix} \underline{I}_\Lambda & \underline{0} \\ \underline{I}_\Lambda & \underline{I}_\Lambda \end{bmatrix}$, with $\underline{I}_\Lambda$ representing the unit of matrix of order Λ, at the matrix of regressors from (4.2.5) in the following way:

$$\begin{bmatrix} \underline{X}_1 & \underline{0} \\ \underline{0} & \underline{X}_2 \end{bmatrix} \begin{bmatrix} \underline{I}_\Lambda & \underline{0} \\ \underline{I}_\Lambda & \underline{I}_\Lambda \end{bmatrix} = \begin{bmatrix} \underline{X}_1 & \underline{0} \\ \underline{X}_2 & \underline{X}_2 \end{bmatrix} \qquad (4.2.7)$$

On the analogy of the use of matrix $\underline{P}_1$ (2.3.12) see page 52, the regression coefficients belonging to (4.2.5) are obtained by pre-multiplying the regression coefficients of (4.2.1a) by the inverse of matrix $\underline{P}$. The result looks as follows:

$$\begin{bmatrix} \hat{\underline{\beta}}_1 \\ \hat{\underline{\gamma}} \end{bmatrix} = \begin{bmatrix} \underline{I}_\Lambda & \underline{0} \\ \underline{I}_\Lambda & \underline{I}_\Lambda \end{bmatrix}^{-1} \begin{bmatrix} \hat{\underline{\beta}}_1 \\ \hat{\underline{\beta}}_2 \end{bmatrix} = \begin{bmatrix} \underline{I}_\Lambda & \underline{0} \\ -\underline{I}_\Lambda & \underline{I}_\Lambda \end{bmatrix} \begin{bmatrix} \hat{\underline{\beta}}_1 \\ \hat{\underline{\beta}}_2 \end{bmatrix} = \begin{bmatrix} \hat{\underline{\beta}}_1 \\ \hat{\underline{\beta}}_2 - \hat{\underline{\beta}}_1 \end{bmatrix}$$

(4.2.8)

The second row of (4.2.8) shows the equality of $\hat{\underline{\gamma}}$ with $(\hat{\underline{\beta}}_2 - \hat{\underline{\beta}}_1)$.
The null hypothesis of Chow's test, viz. $\underline{\beta}_1 = \underline{\beta}_2 = \underline{\beta}_0$, can simply be expressed in a hypothesis relating to $\underline{\gamma}$. If $\underline{\beta}_1 = \underline{\beta}_2$, then $\underline{\beta}_2 - \underline{\beta}_1 = \underline{0}$ also holds, or, expressed in a null hypothesis about $\underline{\gamma}$:

$$\underline{\gamma} = \underline{0} \tag{4.2.9}$$

The null hypothesis (4.2.9) thus invites a test on the significance of $\underline{\gamma}$. The significance test on a number of parameters is mentioned in (2.1.11) on p. 21. To calculate this test statistic, the estimated disturbances of (4.2.5) are needed without hypothesis about $\underline{\gamma}$ being made, as well as the estimated disturbances from (4.2.5) under the null hypothesis $\underline{\gamma} = 0$. The test statistic, as a matter of fact, is composed of a numerator consisting of the difference between the sum of squares of the latter disturbances and the sum of squares of the former disturbances, and a denominator formed by the former sum of squares of disturbances. The estimated disturbances from (4.2.5) are equal to those from (4.2.1a), because $\hat{\underline{\gamma}} = \hat{\underline{\beta}}_2 - \hat{\underline{\beta}}_1$. The disturbances estimated under null hypothesis $\underline{\gamma} = \underline{0}$ are identical with those from (4.1.6). Since Λ parameters are involved and there are $T_1 + T_2 = T$ observations in total, the test statistic reads in this case:

$$F_{[\Lambda, T-2\Lambda]} = \frac{(T-2\Lambda)}{\Lambda} \frac{(\hat{\underline{\varepsilon}}'_{10}\hat{\underline{\varepsilon}}_{10} + \hat{\underline{\varepsilon}}'_{20}\hat{\underline{\varepsilon}}_{20}) - (\hat{\underline{\varepsilon}}'_1\hat{\underline{\varepsilon}}_1 + \hat{\underline{\varepsilon}}'_2\hat{\underline{\varepsilon}}_2)}{(\hat{\underline{\varepsilon}}'_1\hat{\underline{\varepsilon}}_1 + \hat{\underline{\varepsilon}}'_2\hat{\underline{\varepsilon}}_2)} \tag{4.2.10}$$

Now it will be demonstrated that (4.2.10) is identical with the test statistic of Chow's test on p. 133. Chow's test statistic on p. 133 shows the same constant as (4.2.10). The denominator Q_2 from (4.1.10) equals the sum of squares of the estimated disturbances from (4.2.5), because the disturbances of (4.2.5) correspond with those from (4.2.1a) and because these serve for the calculation of Q_2 see (4.1.8). The denominators of (4.2.10) and (4.1.10) are equal, therefore. The numerator of Chow's test statistic is Q_3. The quantity Q_3 can, with the help of (4.1.9), be written as $Q_3 = Q_1 - Q_2$. On comparing the expression $Q_1 - Q_2$ with the numerator of (4.2.10), one perceives that the two expressions are equal to each other, if $Q_1 = \underline{\hat{\varepsilon}}'_{10}\ \underline{\hat{\varepsilon}}_{10} + \underline{\hat{\varepsilon}}'_{20}\ \underline{\hat{\varepsilon}}_{20}$. Since it appears from (4.1.8) that Q_1 is equal to the sum of square of the estimated disturbances under the null hypothesis $\underline{\gamma} = \underline{0}$, the proof of the equality of Chow's test statistic and the significance test on $\underline{\gamma}$ has now been given.

In conclusion it can be remarked that Chow's test is identical with the well-known simultaneous test on the significance of a number of parameters; in particular those parameters are involved which come into being when the transition from one period to another is indicated by a binary dummy variable in the manner as demonstrated in (4.2.4).

The method presented here to test parameters from linear relations for constancy with the help of a binary dummy variable has, however, a number of other advantages over Chow's method. First, the hypothesis stated refers to a number of parameters in one linear relation. For such a null hypothesis, a testing procedure has been developed before. The test gains in accessibility that way, because it can be carried out by means of the basic packet of tests, for which computer programmes are available at almost every research centre where regression analysis is used. Second, the method using a dummy variable can extend Chow's test. The extension consists in the possibility of testing each parameter of $\underline{\gamma}$ separately on significance, whereas Chow's test examines simultaneously the significance of all elements of $\underline{\gamma}$. An economist will be interested in this extension, because he is concerned especially with the individual parameters. Individual parameters indeed describe the economic structure; knowledge of changes in individual parameters may for that reason be more important to him than a collective statement about a number of parameters. The significance of elements of $\underline{\gamma}$ separately is determined by means of the well-known t-test, which is explained in (2.1.18) on page 24.

One can go one step farther by making an attempt to isolate the significant parameters from $\underline{\gamma}$. By this we mean that the binary dummy variable Z_t is maintained only if the relevant parameter of $\underline{\gamma}$ is a significant one. When for instance in the estimation of the parameters from relation (4.2.4) there should occur only one or two significant γ's, these can frequently be isolated by setting the other γ's at zero and estimating anew the parameters of (4.2.4) that are retained. As will be shown in chapter 5, in the cases treated there the parameters from $\underline{\gamma}$ that were significant in the first instance remained so in nearly all cases, which demonstrates the operationality of the isolation procedure. However, the method of isolating the significant elements of $\underline{\gamma}$ is not complete yet. Occasionally it is possible, as will become evident in chapter 5, that a parameter of $\underline{\gamma}$ that was not significant in the first instance, does become significant when the other parameters of $\underline{\gamma}$ are made zero and the relation estimated anew. This phenomenon should be ascribed to the always-present multicollinearity of the economic variables in a regression, leading to shifts in the estimated parameters when variables are omitted or added. The isolation method finally results in the indication and quantification of those parameters of $\underline{\gamma}$

that possess a significant value[11]. Since the other elements of $\underline{\gamma}$ are put equal to zero, there are fewer parameters to be estimated, which implies a gain in degrees of freedom. Moreover, applying the isolation method offers the advantage of a simple interpretation of the estimations, because as an economist one is interested only in the parameters that change at the transition from one period to another.

When the test for constancy of parameters is achieved by means of a binary dummy variable, one linear relation is explicitly made use of. That brings with it the third advantage, which is the one of gaining access to the arsenal of regression methods developed for the treatment of one linear relation; in this connection the treatment of heteroscedasticity and of autocorrelation can be mentioned. When these phenomena occur, the results of a method already in existence, viz. that of generalized least squares, can be applied to test the constancy of the parameters. In Chow's method on the contrary, one would be compelled in such a case to derive the test statistic anew, because the existing test statistic was derived by Chow under a number of assumptions (see p. 132) that exclude the occurrence of each of the two phenomena mentioned above concerning the disturbances.

The alternative approach to the testing problem as regards the constancy of parameters thus in a simple way allows for other assumptions with respect to the disturbances. This third advantage will not be referred to any more in what follows because it does not involve any new applocations of dummy variables.

---

11. Should the situation occur where it is known a priori on economic grounds that one of the γ's deviates from zero, that γ will in most cases be included in the ultimate form of the relation.

Now a generalization of the test for the constancy of parameters will be evolved such that a subdivision of the observation period into more than two subperiods becomes possible. This form of generalization of Chow's test must not be mentioned among the advantages of the dummy variable over Chow's analysis method, because Huang[12] has succeeded in deriving the test for constancy for a subdivision into three periods without using dummy variables.

An advantage of the formulation with dummy variables in (4.2.5) is, however, the possibility to introduce with very little trouble any other subdivision than a division into two parts, and to test the relevant parameters for their constancy. As an example a division into four subperiods will be discussed.

With economic time series consisting of observations per quarter the question arises whether one and the same set of parameters satisfactorily represents the structure of the relation between the relevant economic variables for each of the quarters separately. The question leads to a test for equality carried out on the sets of parameters belonging to each of the quarters. With individual treatment of each of the four quarters, the following linear relations prevail:

$$\begin{aligned}
\underline{Y}_{I} &= \underline{X}_{I}\underline{\beta}_{I} + \underline{\varepsilon}_{I}\\
\underline{Y}_{II} &= \underline{X}_{II}\underline{\beta}_{II} + \underline{\varepsilon}_{II}\\
\underline{Y}_{III} &= \underline{X}_{III}\underline{\beta}_{III} + \underline{\varepsilon}_{III}\\
\underline{Y}_{IV} &= \underline{X}_{IV}\underline{\beta}_{IV} + \underline{\varepsilon}_{IV}
\end{aligned} \qquad (4.2.11)$$

or

12. D.S. Huang, Regression and econometric methods, New York 1970, pp. 108 ff.

$$\begin{bmatrix} \underline{Y}_{I} \\ \underline{Y}_{II} \\ \underline{Y}_{III} \\ \underline{Y}_{IV} \end{bmatrix} = \begin{bmatrix} \underline{X}_{I} & \underline{0} & \underline{0} & \underline{0} \\ \underline{0} & \underline{X}_{II} & \underline{0} & \underline{0} \\ \underline{0} & \underline{0} & \underline{X}_{III} & \underline{0} \\ \underline{0} & \underline{0} & \underline{0} & \underline{X}_{IV} \end{bmatrix} \begin{bmatrix} \underline{\beta}_{I} \\ \underline{\beta}_{II} \\ \underline{\beta}_{III} \\ \underline{\beta}_{IV} \end{bmatrix} + \begin{bmatrix} \underline{\varepsilon}_{I} \\ \underline{\varepsilon}_{II} \\ \underline{\varepsilon}_{III} \\ \underline{\varepsilon}_{IV} \end{bmatrix} \qquad (4.2.12)$$

where:

$\underline{Y}_{I}$: observations of the regressand from the first quarter

$\underline{X}_{I}$: observations of the regressors for the first quarter (the number of regressors is Λ)

$\underline{\beta}_{I}$: vector with parameters pertaining to the first quarter

$\underline{\varepsilon}_{I}$: disturbances pertaining to the observations of the first quarter.

The same, mutatis mutandis, for indices II, III and IV. The test for constancy of the parameters can be formulated as a null hypothesis as follows:

$$\underline{\beta}_{I} = \underline{\beta}_{II} = \underline{\beta}_{III} = \underline{\beta}_{IV} = \underline{\beta}_{0} \qquad (4.2.13)$$

Under the null hypothesis (4.2.13), relation (4.2.12) passes into[13]:

$$\begin{bmatrix} \underline{Y}_{I} \\ \underline{Y}_{II} \\ \underline{Y}_{III} \\ \underline{Y}_{IV} \end{bmatrix} = \begin{bmatrix} \underline{X}_{I} \\ \underline{X}_{II} \\ \underline{X}_{III} \\ \underline{X}_{IV} \end{bmatrix} \underline{\beta}_{0} + \begin{bmatrix} \underline{\varepsilon}_{I0} \\ \underline{\varepsilon}_{II0} \\ \underline{\varepsilon}_{III0} \\ \underline{\varepsilon}_{IV0} \end{bmatrix} \qquad (4.2.14)$$

13. The disturbance vectors are provided with index 0 to distinguish them from the disturbance vectors in (4.2.12).

The testing problem described above is tackled as follows with the help of binary dummy variables. On the analogy of the relation between (4.2.1a) and the alternative formulation (4.2.5), (4.2.12) can be written with the help of binary dummy variables as follows:

$$\begin{bmatrix} \underline{Y}_I \\ \underline{Y}_{II} \\ \underline{Y}_{III} \\ \underline{Y}_{IV} \end{bmatrix} = \begin{bmatrix} \underline{X}_I & \underline{0} & \underline{0} & \underline{0} \\ \underline{X}_{II} & \underline{X}_{II} & \underline{0} & \underline{0} \\ \underline{X}_{III} & \underline{0} & \underline{X}_{III} & \underline{0} \\ \underline{X}_{IV} & \underline{0} & \underline{0} & \underline{X}_{IV} \end{bmatrix} \begin{bmatrix} \underline{\beta}_I \\ \underline{\beta}_{II} - \underline{\beta}_I \\ \underline{\beta}_{III} - \underline{\beta}_I \\ \underline{\beta}_{IV} \quad \underline{\beta}_I \end{bmatrix} + \begin{bmatrix} \underline{\varepsilon}_I \\ \underline{\varepsilon}_{II} \\ \underline{\varepsilon}_{III} \\ \underline{\varepsilon}_{IV} \end{bmatrix} \tag{4.2.15}$$

The null hypothesis (4.2.13) can be reduced to the following null hypothesis relating to the parameters of (4.2.15):

$$\underline{\beta}_{II} - \underline{\beta}_I = \underline{\beta}_{III} - \underline{\beta}_I = \underline{\beta}_{IV} - \underline{\beta}_I = \underline{0} \tag{4.2.16}$$

Evidently, null hypothesis (4.2.13) can be tested by testing the significance of the last 3Λ parameters of (4.2.15). The corresponding test statistic is obtained as explained in (2.1.11) on page 21. With T observations and 4Λ parameters, out of which 3Λ are to be tested on significance, one obtains a test statistic distributed according to $F_{[3\Lambda, T-4\Lambda]}$. The numerator of this test statistic - not counting the constant - consists in the sum of the squares of the estimated disturbances under the null hypothesis (see (4.2.14)) less the sum of squares of the estimated disturbances of the relation (4.2.15), or, if preferred (4.2.12). The denominator of the test statistic consists in the latter sum of squares. In a formula, the test statistic reads:

$$F_{[3\Lambda, T-4\Lambda]} = \frac{(T-4\Lambda)}{3\Lambda} \; \frac{\sum_{i=I}^{IV} \hat{\underline{\varepsilon}}'_{i0}\, \hat{\underline{\varepsilon}}_{i0} - \sum_{i=I}^{IV} \hat{\underline{\varepsilon}}'_{i}\, \hat{\underline{\varepsilon}}_{i}}{\sum_{i=I}^{IV} \hat{\underline{\varepsilon}}'_{i}\, \hat{\underline{\varepsilon}}_{i}} \qquad (4.2.17)$$

In (4.2.17) an alternative is presented for the method Klein[14] et al. have used to test the constancy of parameters from quarter to quarter. Their method, characterized as deficient and subjective, can thus be replaced by a test that is founded on statistic theory.

It remains to be demonstrated that the method of testing on constancy set out in this section, can be applied if only a subset of the parameters should be tested on constancy; Chow's approach can be found on pages 134-137. We start from relation (4.1.11a) which reads:

$$\begin{bmatrix} \underline{Y}_1 \\ \underline{Y}_2 \end{bmatrix} = \begin{bmatrix} \underline{X}_{1a} & \underline{0} & \underline{X}_{1b} & \underline{0} \\ \underline{0} & \underline{X}_{2a} & \underline{0} & \underline{X}_{2b} \end{bmatrix} \begin{bmatrix} \underline{\beta}_{1a} \\ \underline{\beta}_{2a} \\ \underline{\beta}_{1b} \\ \underline{\beta}_{2b} \end{bmatrix} + \begin{bmatrix} \underline{\varepsilon}_1 \\ \underline{\varepsilon}_2 \end{bmatrix} \qquad (4.2.18)$$

It is known a priori that $\underline{\beta}_{1a}$ and $\underline{\beta}_{2a}$ are not equal. The equality of $\underline{\beta}_{1b}$ with $\underline{\beta}_{2b}$ is to be examined, which can be reproduced as follows in a null hypothesis:

$$\underline{\beta}_{2b} - \underline{\beta}_{1b} = \underline{0} \qquad (4.2.19)$$

Analogously with the situation in the previous cases of this section, the approach with a binary dummy variable leads to the relation:

$$\begin{bmatrix} \underline{Y}_1 \\ \underline{Y}_2 \end{bmatrix} = \begin{bmatrix} \underline{X}_{1a} & \underline{0} & \underline{X}_{1b} & \underline{0} \\ \underline{0} & \underline{X}_{2a} & \underline{X}_{2b} & X_{2b} \end{bmatrix} \begin{bmatrix} \underline{\beta}_{1a} \\ \underline{\beta}_{2a} \\ \underline{\beta}_{1b} \\ \underline{\beta}_{2b} - \underline{\beta}_{1b} \end{bmatrix} + \begin{bmatrix} \underline{\varepsilon}_1 \\ \underline{\varepsilon}_2 \end{bmatrix} \qquad (4.2.20)$$

14. See for source and description of the method pp. 74 and 75.

Null hypothesis (4.2.19) is tested in terms of (4.2.20) by checking the significance of the last vector of parameters, in this case $\underline{\beta}_{2b} - \underline{\beta}_{1b}$. The proof that by following the above method one obtains the same test statistic that Chow got, see formula (4.1.16), is not developed here, because it does not reveal any new points of view. Concluding, one may state that the method evolved in this section to test the constancy of parameters possesses wider possibilities than Chow's, and is, therefore preferable.

Now we shall briefly[15] go into the relation between testing the constancy of parameters as described above, and analysis of covariance. A situation explained by Malinvaud[16] will serve as an example. He starts from the simple linear relation

$$Y_{ij} = \beta_{0j} + \beta_{1j}X_{ij} + \varepsilon_{ij} \tag{4.2.21}$$

where

- Y_{ij} : yearly expenditure on footwaer per consumption unit in household i (i = 1,2,...,N); household i belongs to the socio-professional class j (j = 1,2,...,p)
- X_{ij} : yearly total expenses per consumption unit in household i, the household belongs to the socio-professional class j
- β_{0j} and β_{1j}: parameters that are dependent on the class to which the relevant household belongs
- ε_{ij} : the disturbance; for which the assumptions 1, 2 and 4 of pp. 18-19 hold.

In the setting of analysis of covariance Malinvaud arriv-

15. See for a more extensive treatment Johnston, ibid. p. 8, pp. 192 ff.
16. Ibid. p. 3, pp. 233-237.

es at the following table:

Source of variation	Degrees of freedom	Quadratic form
Within classes with unequal effects	N - 2p	$A = \sum_{j=1}^{p} \Sigma_i^j (Y_{ij}-\bar{Y}_{.j})^2 - \sum_{j=1}^{p} \hat{\beta}_{1j} \Sigma_i^j (Y_{ij}-\bar{Y}_{.j})(X_{ij}-\bar{X}_{.j})$
Differences among the β_{1j}	p - 1	$B = \sum_{j=1}^{p} (\hat{\beta}_{1j}-\hat{\beta}_1) \Sigma_i^j (Y_{ij}-\bar{Y}_{.j})(X_{ij}-X_{.j})$
Within classes with equal effects	N-p-1	$C = \sum_{j=1}^{p} \Sigma_i^j (Y_{ij}-\bar{Y}_{.j})^2 - \hat{\beta}_1 \sum_{j=1}^{p} \Sigma_i^j (Y_{ij}-\bar{Y}_{.j})(X_{ij}-\bar{X}_{.j})$

The meaning of the symbols used in the table is:

$\hat{\beta}_{1j}$: least-squares estimator for β_{1j}, estimated on the basis of the observation pairs from class j

$\hat{\beta}_1$: least-squares estimator for β_1 under the hypothesis $\beta_{11} = \beta_{12} = \ldots = \beta_{1p} = \beta_1$

Σ_i^j : sum taken over i as far as the elements belong to class j

$\bar{Y}_{.j}, \bar{X}_{.j}$: means of variables Y and X, respectively, calculated for class j

To test the hypothesis of equality of the β_{1j}'s, Malinvaud used the following statistic distributed according to $F_{[p-1,N-2p]}$:

$$Q = \frac{N-2p}{p-1} \quad \frac{C-A}{A} \tag{4.2.22}$$

With the help of binary dummy variables the relation reproduced in (4.2.21) can be rewritten. To that end the following set of binary dummy variables is chosen to express the subdivision into p classes:

$$\begin{aligned} Z_{i2} &= 1 \text{ for observations in class 2} \\ &= 0 \text{ for all other observations} \end{aligned}$$

$$\begin{aligned} Z_{i3} &= 1 \text{ for observations in class 3} \\ &= 0 \text{ for all other observations} \end{aligned}$$

and so on, the last of the p-1 dummy variables reading

$$\begin{aligned} Z_{ip} &= 1 \text{ for observations in class p} \\ &= 0 \text{ for all other observations.} \end{aligned}$$

Relation (4.2.21) can now be written as

$$\begin{aligned} Y_{ij} = {} & \beta_{01} + \gamma_{02} Z_{i2} + \gamma_{03} Z_{i3} + \ldots + \gamma_{0p} Z_{ip} + \\ & + \beta_{11} X_{ij} + \gamma_{12} Z_{i2} X_{ij} + \gamma_{13} Z_{i3} X_{ij} + \ldots + \gamma_{1p} Z_{ip} X_{ij} + \varepsilon_{ij} \end{aligned} \tag{4.2.23}$$

where

$$\begin{aligned} \gamma_{02} &= \beta_{02} - \beta_{01}, \quad \gamma_{03} = \beta_{03} - \beta_{01} \quad \text{etc.} \\ \gamma_{12} &= \beta_{12} - \beta_{11}, \quad \gamma_{13} = \beta_{13} - \beta_{11} \quad \text{etc.} \end{aligned}$$

If (4.2.23) is written in matrix notation for all observations, there results:

$$
\begin{bmatrix} \underline{Y}_1 \\ \underline{Y}_2 \\ \underline{Y}_3 \\ \cdot \\ \cdot \\ \cdot \\ \underline{Y}_p \end{bmatrix}
=
\begin{bmatrix}
\underline{1}_1 & \underline{0} & \underline{0} & \cdots & \underline{0} & \underline{X}_1 & \underline{0} & \underline{0} & \cdots & \underline{0} \\
\underline{1}_2 & \underline{1}_2 & \underline{0} & \cdots & \underline{0} & \underline{X}_2 & \underline{X}_2 & \underline{0} & \cdots & \underline{0} \\
\underline{1}_3 & \underline{0} & \underline{1}_3 & \cdots & \underline{0} & \underline{X}_3 & \underline{0} & \underline{X}_3 & \cdots & \underline{0} \\
\cdot & \cdot & \ddots & & \cdot & \cdot & \cdot & \ddots & & \cdot \\
\cdot & \cdot & \cdot & \ddots & \cdot & \cdot & \cdot & \cdot & \ddots & \cdot \\
\cdot & \cdot & \cdot & & \ddots & \cdot & \cdot & \cdot & & \ddots \\
\underline{1}_p & \underline{0} & \underline{0} & \cdots & \underline{1}_p & \underline{X}_p & \underline{0} & \underline{0} & \cdots & \underline{X}_p
\end{bmatrix}
\begin{bmatrix} \beta_{01} \\ \gamma_{02} \\ \gamma_{03} \\ \cdot \\ \cdot \\ \cdot \\ \gamma_{0p} \\ \beta_{11} \\ \gamma_{12} \\ \gamma_{13} \\ \cdot \\ \cdot \\ \cdot \\ \gamma_{1p} \end{bmatrix}
+
\begin{bmatrix} \underline{\varepsilon}_1 \\ \underline{\varepsilon}_2 \\ \underline{\varepsilon}_3 \\ \cdot \\ \cdot \\ \cdot \\ \underline{\varepsilon}_p \end{bmatrix}
\qquad (4.2.24)
$$

where

$\underline{Y}_1$: the vector of the regressand values of class 1

$\underline{1}_1$: the vector with as many elements equal to 1 as there are observations of class 1

$\underline{X}_1$: the vector of regressand values of class 1

$\underline{\varepsilon}_1$: the vector of disturbances for class 1

Similarly, mutatis mutandis, for indices 2 through p. The equality hypothesis of the β_{1j}'s can be reformulated as a null hypothesis for the last p-1 parameters of the linear relation (4.2.24), viz.

$$
\gamma_{12} = \gamma_{13} = \ldots = \gamma_{1p} = 0 \qquad (4.2.25)
$$

The test statistic mentioned in (2.1.11) see page 21, which has been developed for testing the significance of a number of parameters, is suitable for this purpose. For the above case the test statistic reads:

$$Q_1 = \frac{N-2p}{p-1} \frac{\hat{\underline{\varepsilon}}_0'\hat{\underline{\varepsilon}}_0 - \hat{\underline{\varepsilon}}'\hat{\underline{\varepsilon}}}{\hat{\underline{\varepsilon}}'\hat{\underline{\varepsilon}}} \qquad (4.2.26)$$

in which $\hat{\underline{\varepsilon}}_0$ represents the least-squares estimator for the disturbance vector of (4.2.24) under hypothesis (4.2.25), and $\hat{\underline{\varepsilon}}$ the least-squares estimator for the disturbance vector in (4.2.24), no restrictions being imposed. Test statistic Q_1 follows the distribution $F_{(p-1,N-2p)}$. It can simply be checked that test statistics (4.2.22) and (4.2.26) are identical.
Thus it has been demonstrated that analysis of covariance and regression analysis with dummy variables are essentially equivalent techniques.

4.3. THE POWER OF THE TEST FOR CONSTANCY OF THE PARAMETERS IN LINEAR RELATIONS

The usefulness of a test is closely linked up with its probability of justly accepting an alternative hypothesis; in this connection one speaks of the power of a test. A discussion of the power of the test is, therefore, warranted on all accounts.
No calculations of the power of Chow's test have been published so far. In this section the theoretical derivation of the power of Chow's test will be given. In chapter 5 some numerical results will be presented.
It was shown that Chow's test statistic can be reformulated in the form of a significance test on the last Λ parameters of (4.2.5). That finding will now be used to assess the power of Chow's test, the power of the significance test for $\underline{\gamma}$ having already been theoretically derived, see p. 22. As to the general significance test

on $\underline{\gamma}$ from the relation (4.2.5) it appears that the relevant test statistic (4.2.10), follows a non-central $F_{[\Lambda,T-2\Lambda,\lambda]}$-distribution, λ representing the so-called non-centrality parameter. This reads

$$\lambda = \frac{(\underline{\gamma}-\underline{\gamma}_0)'B(\underline{\gamma}-\underline{\gamma}_0)}{2\sigma^2} \tag{4.3.1}$$

where

$\underline{\gamma}_0$: the vector of Λ parameters that are equal to the corresponding parameters of $\underline{\gamma}$ under the null hypothesis;

$\underline{B}$: a positively definite matrix of the shape:

$$\underline{B} = \underline{X}_2'\underline{X}_2 - \underline{X}_2'\underline{X}_2(\underline{X}_1'\underline{X}_1 + \underline{X}_2'\underline{X}_2)^{-1}\underline{X}_2'\underline{X}_2 \tag{4.3.2}$$

Under null hypothesis (4.2.9), i.e. $\underline{\gamma} = 0$, λ has the value zero and the test statistic follows a central distribution according to $F_{[\Lambda,T-2\Lambda]}$. If as an alternative hypothesis for $\underline{\gamma}$ it is proposed, for example, that $\underline{\gamma} = \underline{\gamma}_1$, then test statistic (4.2.10), and hence also Chow's test statistic (4.1.10) follow a non-central F-distribution with parameter ($\underline{\gamma}_0$ equals $\underline{0}$)

$$\lambda_1 = \frac{\underline{\gamma}_1'B\underline{\gamma}_1}{2\sigma^2} \tag{4.3.3}$$

The power of Chow's test, to be denoted by $p(\underline{\gamma}_1)$, is obtained next by calculating at a chosen significance level α the integral over the critical region of the distribution of the test statistic under the alternative hypothesis. The power is expressed as a formula, with one-sided testing[17], thus:

---

17. Compare (2.1.11).

$$p(\underline{\gamma}_1) = \int_{F_{\alpha[\Lambda,T-2\Lambda]}}^{\infty} F_{[\Lambda,T-2\Lambda,\lambda_1]}\, df \qquad (4.3.4)$$

where

$F_{\alpha[\Lambda,T-2\Lambda]}$: the table value at significance level α for an F-distribution with Λ degrees of freedom in the numerator and $T-2\Lambda$ degrees of freedom in the denominator.

The possibility of determining in a simple way the power of the constancy test for parameters reinforces the conclusion reached at the end of section 4.2, viz., that testing parameters for constancy by means of dummy variables offers wider possibilities than Chow's approach.

5. Application: investigation into structural differences in the Netherlands economy by means of an econometric model for the pre-war and post-war years

5.1. INTRODUCTION

The Central Planning Bureau (C.P.B.) has since 1945 been concerned with building econometric models for the Netherlands economy, for the purpose of making short-term forecasts as well as studying problems of an economic-political nature. An extensive model, which is known by its code number 63-D, was published for the first time in 1964[1]. Model 63-D, which serves as an example in this chapter, played until 1970[2] an important part in the forecasts the C.P.B. publishes every year in the framework of the Central Economic Plan. Econometric model 63-D, which is reproduced in extenso in the Appendix to this chapter, consists of 38 equations, in which approximately 100 variables occur. The equations of the model are distinguished by the C.P.B. into reaction equations and definitional equations; of the first kind the model counts 13, the other 25 are definitional equations. In total the 13 reaction equations contain 77 parameters characterizing the economic structure of the Netherlands. The investigation in this chapter focuses on the constancy over time of these parameters.
The estimations of the parameters of the model are, in the procedure followed by the C.P.B. based on series of

---

1. See note 21 on p. 8.
2. Meanwhile the C.P.B. has designed a variant of model 63-D, viz. 69-C; this new model has published in the Central Economic Plan 1971.

observations of economic magnitudes going back to the year 1923. The observations of the variables are in terms of relative first differences; the tth observation of e.g., variable y, is obtained as follows:

$$y_t = 100 \frac{Y_t - Y_{t-1}}{Y_{t-1}} \qquad (5.1.1)$$

The observation material falls into two categories, viz. a pre-war period with observations from 1923 through 1939, and a post-war period, beginning with the observations of the year 1949. The character of the two periods is quite different.

Economic life in the pre-war period was characterized among other things by severe cyclical fluctuations, which reached a nadir during the depression of the thirties. The post-war period, on the contrary, shows an almost continuous upward movement in the economy. Under those circumstances recessions have, generally speaking, the character of a decrease in growth; appreciable declines in economic activity have not occurred. In view of the differences[3] in economic climate between the two periods the question is warranted whether it is justified to operate in the model, in calculations relevant to the whole period, parameters that are supposed to have remained constant since 1923. For the post-war period as such, too, one may ask if the parameters have remained constant. It could be tested, for instance, whether the rate of inflation appreciably affects the value of the parameters in the model. Because this chapter is meant to serve an an illustration of the test for constancy of parameters, the contents will be limited to a comparison

---

3. See also J.H.F. Schilderinck, Een ekonometrisch model van de Nederlandse ekonomie, Ph.D. defended at the Catholic University at Tilburg, 1970, pp. 22 ff.

of the pre- and post-war periods.
As was explained in chapter 4, the constancy of the parameters can be verified with Chow's test if a collective verdict on all the parameters in an equation is required, and by means of dummy variables, if on top of that one wants to test the constancy of each parameter separately. Both methods are applied in section 5.2. Attention is directed first and foremost to the latter method, because that one leads to the indication of those among the parameters used by the C.P.B. about which the hypothesis of constancy must be rejected. When each reaction equation is examined individually[4], it appears that for 14 per cent of the total number of parameters that are supposed constant by the C.P.B., the hypothesis of constancy must be rejected. In other words, 14 per cent of the parameters has for the post-war years a value that differs significantly from the pre-war value. The probability of obtaining this large number of non-constant parameters under the hypothesis of constancy is very small, so that the conclusion must be that it is unjustified to work with coefficients calculated for the whole observation period.
Discarding the constancy of a number of parameters from the C.P.B.-model 63-D need not mean that the pre-war data are useless for the estimation of parameters that are relevant in the post-war period. As long as the C.P.B. has chosen, for each equation of model 63-D, on the one hand variables corresponding with the specification of a model tailored for pre-war years, and on the other hand variables corresponding with the specification of a model

4. The question whether in this investigation the simultaneity of the model can and should be taken into account will be elaborated upon in the course of this section.

tailored for post-war years, and if these specifications are identical the inequality of corresponding parameters does not matter. For, using binary dummy variables, one can indicate changes in parameters; these changes will be quantified below with the help of the method discussed on pp. 129 and 130. About the degree of reality of the assumption regarding the similarity between the specifications for each individual period a few remarks will be made now.

In designing a model for a specific period, the designer will aim at including such economic variables - always within economic theory - as have non-zero parameters. The investigation in this chapter with respect to the constancy of the parameters will be accompanied by the calculation of pre-war coefficients as well as the difference between corresponding post-war and pre-war coefficients (see (4.2.8)). From the figures thus obtained it is possible to derive the coefficients for the two periods separately. Among the parameters of the pre-war period, slightly more than 60 per cent have values that do not deviate significantly from zero; for the post-war period this percentage is just over 50. From the height of the two percentages the conclusion seems obvious that the specification of model 63-D fits neither of the two periods. That nevertheless the parameters of model 63-D, which have been estimated on the basis of the observations of the whole period, lead only in just over 20 per cent of the cases considered to values not deviating from zero should be ascribed to two factors:

1. the lenght of the time series used, leading to a greater number of degrees of freedom, and
2. the occurence of clusters owing to the combination of the two periods: one cluster consists of the many

positive relative changes in the variables after the war, another of the many negative relative changes before the war [5].

The above objections to model 63-D lead to the construction of a model accounting for the specific character of the post-war period.
The test on the constancy of the parameters of model 63-D will be carried out in section 5.2, on the one hand with the help of the test developed by Chow discussed in section 4.1, on the other hand by the method developed in section 4.2, in which dummy variables are used. In chapter 4 the constancy test on parameters from one individual relation have been dealt with. Contrary to the situation in chapter 4, model 63-D consists of a large number of simultaneous relations. First, it will be explained why ample attention will be given to the results of a non-simultaneous estimating method, viz. that of ordinary least-squares (OLS) rather than to the results of the much-used two-stage least-squares method (2SLS).
It is known[6] that application of the OLS estimating method to individual relations of a simultaneous system of equations results in non-consistent estimators for the parameters. Applying the 2SLS method does yield consistent estimators. However, in many cases one has to pay dearly for the property of consistency, in the form of larger standard deviations[7]. Moreover, in practical

---

5. Two clusters in the observation material determine to a great extent the regression results.
6. See, e.g., Johnston, ibid. p. 8, pp. 341 ff.
7. See the review of simulation studies by Johnston, ibid. p. 8, pp. 408 ff. Compare also the standard deviations of the coefficients of model 63-D for both estimating methods on pp. 178 ff.

applications consistency is not all that important, because one often has to work with small examples. All the same, the simulation studies mentioned by Johnston seem to suggest that for small models the method of 2SLS is to be preferred to the OLS method.

With larger models, like model 63-D, another difficulty arises which renders a simultaneous estimating method as 2SLS less attractive. Because a choice has to be made among the abundance of pre-determined variables of the model to select those that will be used in the first stage of the estimating procedure[8], inevitably a subjective element slinks into the application and into the results of the 2SLS method. One has the predetermined variables of the model and/or the principal components of these predetermined variables to choose from. The ultimate choice can, therefore, consist of: first, a number of predetermined variables only, second, a number of principal components only, and third, a number of principal components, completed by some pre-determined variables. That the choice of the number of variables to be included as well as the composition of the choice made, affects the estimation results, may be illustrated with an example.

Equation 10 of the C.P.B. model 63-D (see p. 181) has to

8. Theil (H. Theil, Economics Forecasts and policy, Amsterdam 1958, p. 342 ff.) prints this remark. He suggests replacement of the abundance of predetermined variables with a smaller number of principal components. With T. Kloek and L.B. Mennes, Simultaneous equations estimation based on principal components of predetermined variables, Econometrica, Vol. 28 (1960), No. 1, pp. 45-60, Theil's suggestion is elaborated. The 'Method IV' proposed by Kloek and Mennes is used in chapter 5 of this study. F.M. Fisher, in his article, The choice of instrumental variables in the estimation of economy-wide econometric models, International Economic Review, Vol. 6 (1965), No. 3, pp. 245-274, discusses some of these methods.

that end been estimated with five different selections of sets of pre-determined variables and of principal components to be used in the first stage. The result looks as follows; for the standard deviations an average of the five estimated standard deviations has been calculated because the differences obtained were small.

	variable		
	H	p_m	p_i
Highest value of the coefficient found	.54	.44	.38
Lowest value of the coefficient found	.23	.39	.28
Average standard deviation	.20	.05	.08

From the above table it appears that the highest and lowest values of the coefficients deviate from each other by once or more times the standard deviation. It is not at all an exaggeration to use this result as a rule of thumb for the whole model. The conclusion seems warranted that the subjective element hampers interpretation and operation of estimations obtained with 2SLS. This objection will become even stronger when, moreover, the right is reserved to make a choice from the sumultaneous estimating methods available, and when some equations are estimated by one, others by the other method. If two estimating methods are used side by side, as is done by the C.P.B. then a two-fold choice is available, viz., first from two estimating methods, and second, from a number of sets of pre-determined variables or functions of these to be used in the first stage. The estimated values of the parameters become thus highly subjective, which reduces the constancy test to a less meaningful activity.

Yet another reason, on another level, can be put forward

for the exclusive use of ordinary least-squares in a constancy test. Usually such a test will be carried out while the model is under construction. The specification of an equation is also subject to change if one of the parameters should appear not to be constant. In a stage in which the ultimate specification has not been fixed yet, simultaneous estimating techniques cannot be applied, because they precisely call for knowledge of the definite specification.

If isolation and quantification of the significant discontinuous changes by means of the two-stage least-squares method should nevertheless be wanted, then the method described in section 4.2 would have to be adapted to the purpose. To that effect one could proceed as follows. In the first stage the least-squares estimations for the 'explanatory' endogenous variables of an equation of the model are calculated. These 'cleaned' variables are substituted for the original ones. Next, the technique from section 4.2 is applied in the second stage to isolate and quantify the significant discontinuous changes in the parameters.

The method explained in the previous paragraph will, however, not be carried out, on the earlier-mentioned motive of subjectivity in the results that are obtained by, for example, the estimation method of 2SLS. To avoid subjectivity, in section 5.2 the significant changes in the parameters will be isolated and quantified by means of the ordinary least-squares method. Next, for each significant change a binary dummy variable will be introduced. For the new specification of the C.P.B. model achieved in this way, least-squares estimations are obtained. For good measure, the parameters of the thus accomplished new specification of the C.P.B. model will also be estimated by means of the two-stage least-squares method. The variant of this estimation method chosen is that developed by Kloek and Mennes[9].

Section 5.3 contains the results of a number of calculations of the power of Chow's test. The null hypothesis of this test is that the parameters of the pre-war period are equal to the corresponding parameters of the post-war period. In (4.1.5) on page 132 this null hypothesis is reproduced by $\underline{\beta}_1 = \underline{\beta}_2 = \underline{\beta}_0$, where $\underline{\beta}_1$ represents the vector of pre-war parameters and $\underline{\beta}_2$ the post-war vector. As appears from (4.2.6) and (4.2.9), see pp. 141 and 142, the null hypothesis can also be denoted by $\underline{\gamma} = \underline{\beta}_2 - \underline{\beta}_1 = \underline{0}$. As an alternative hypothesis, in principle any vector of numbers can be chosen for $\underline{\gamma}$. Since $\underline{\gamma}$ represents the difference between the pre-war and post-war parameters, it is possible to build on the corresponding estimates a number of alternative hypothesis. Pursuing this idea for each of the 13 reaction equations, three alternative hypothesis will be evolved in section 5.3.

5.2. RESULTS OF THE INVESTIGATION INTO THE CONSTANCY OF THE PARAMETERS OF C.P.B.-MODEL 63-D

The investigation to be discussed is based on the specification of the C.P.B. model as it is described in the appendix to this chapter. The observation period covers 16 pre-war and 17 post-war years. In the first instance, a binary dummy variable was included for every parameter in the manner explained on page 140. The introduction of a binary dummy variable for each of the 74 parameters brings the total number of parameters to be estimated to

9. Ibid. p. 162, pp. 53 ff. The number of principal components to be used, which has not been fixed in Kloek and Mennes, has been put in this study at 8 for each reaction equation.

148.

Before the estimation of these 148 parameters was taken in hand, it was verified, for the observations underlying model 63-D, whether or not the disturbances calculated for the whole observation period on the basis of the estimates of the 74 parameters, showed any autocorrelation[10]. The hypothesis of absence of autocorrelation can be tested with the test statistic developed by Durbin and Watson. As is evident from the results of this test statistic (see p. 178 ff.), on a significance level of 5 per cent one cannot reject the hypothesis of absence of autocorrelation. Besides it has been tested whether the assumption of absence of heteroscedasticity holds[11], on the basis of disturbances calculated with the help of relations estimated for the pre-war and post-war periods separately. In the latter test, contrary to the former, it was necessary to estimate the disturbances for the individual periods, because for a test on heteroscedasticity one needs the variances of the disturbances of the separate periods. With the F-test it was examined for each of the 13 reaction equations whether the estimated variances of pre-war disturbances deviate significantly from the corresponding variances of the post-war disturbances. On a significance level of 5 per cent, 2 out of the 13 test statistics are significant, a result however that has a probability of over 5 per cent of occurring. For, if a test on a 5 per cent level is carried out 13 times, the probability of finding twice or more a significant test statistic is 14 per cent.

10. The assumption of absence of autocorrelation as well as that of absence of heteroscedasticity are considered applicable to the method developed in section 4.2, the first, second and fourth assumption of pp. 18 and 19 being used.
11. This probability has been calculated by means of the Poission distribution with parameter 0.65.

Since in this chapter a significance level of 5 per cent is generally adopted, the hypothesis of absence of heteroscedasticity cannot be rejected on the ground of this result.
Next, for each of the 13 behavioural equations of the model the pre-war parameters were estimated by the method of ordinary least-squares, as well as their discontinuous changes at the transition from the pre-war to the post-war period. In the symbolism of section 4.2 that implies estimating $\underline{\beta}_1$ and $\underline{\gamma} = \underline{\beta}_2 - \underline{\beta}_1$. The least-squares estimation of all the 74 parameters representing the discontinuous changes (the elements of $\underline{\gamma}$) produce 7 coefficients that are significant on a 5 per cent level, two of which are significant on a level of 1 per cent. These 7 coefficients belong to, specifically: variable c^r_{-1} in equation 2, v_a in equation 6, $H_{-\frac{1}{2}}$ and $T'_{K-\frac{1}{3}}$ in equation 9, H in equation 10, the constant-term variable in equation 11 and H in equation 12. These results will be analysed after the isolation of the significant discontinuous changes has been completed.
The simultaneous test of Chow, which, as is known, refers to all parameters of an equation, shows significant values for equations 6 and 9. According to reasoning similar to that for the 13 tests on heteroscedasticity, on the ground of the results of Chow's test the hypothesis of constancy of the parameters may not be rejected. The estimates thus obtained of the pre-war parameters and of the differences between pre-war and post-war parameters, can serve to derive the estimates for the post-war parameters. The post-war coefficients are obtained simply by taking the sum of corresponsing elements of $\hat{\underline{\beta}}_1$ and $\hat{\underline{\gamma}} = \hat{\underline{\beta}}_2 - \hat{\underline{\beta}}_1$[12]. When the pre-war and post-war coefficients themselves (not their differences) are

12. See for footnote 12, page 168.

examined as to significance, it appears that 47, or 63 per cent, of the pre-war parameters, and 40, or 54 per cent, of the post-war parameters do not have a value that deviates significantly from zero.

Finally it has been examined if it was possible to isolate the non-constant parameters. With the dummy variable method it is possible to obtain estimated differences between pre- and post-war values for all 74 parameters of the model. Among them are a large number of non-significant differences. In model 63-D, as can be inferred from the above results, in the first instance 74-7 = 67 difference parameters are not significant. It has already been mentioned as an advantage of the approach by dummy variables (see pp. 143-144) that it will be possible to indicate among the large number of estimated changes in corresponding post- and pre-war parameters a limited number as significant, and to put the difference parameters relevant to the others equal to zero. The operationality of this new method is now being examined. It is not sufficient, however, to introduce binary dummy variables for the significant parameters expressing differences between pre- and post-war parameters, and then to estimate once more, under the restriction that the difference parameters relevant to differences found non-significant in the first instance have the value zero. As a matter of fact the method referred to implies the possibility that, some difference parameters being put at zero, a number of the difference parameters found non-significant in the first instance may as yet become significant. For all non-significant parameters that have an estimated value of more than once the relevant

---

12. An explanation of this method and a derivation of the relevant standard deviations is found, among others, in Goldberger, ibid. p. 7, pp. 167 and 168.

standard deviation, it has been attempted to get a significant parameter after all, by putting at zero the other difference parameters from the equation concerned. Only in three cases has the attempt been successful, so that the list of significant difference parameters has been extended to 10 items. The results of estimations carried out for model 63-D after changing the specification for these 10 significant discontinuous changes in the parameters, are given at the end of this section under the headings A. For the specification given under A the parameters have also been estimated by the 2SLS method, for the sake of completeness; the results are given under the headings B. When judging the B-results one should bear in mind that the isolation of significant differences had not been carried out by this method, in view of the objections to this method in this case (see pp. 161-163).

The number of 10 differences that are significant on a 5 per cent level - 2 of them being significant on a level of 1 per cent - has under the hypothesis of constancy a very small probability of occurring for when a significance test on a 5 per cent level is carried out 74 times, the number of 'significant' outcomes to be ascribed to chance follows a Poisson-distribution with parameter $\lambda = 3.7$. In this case there is a probability of only 0.5 per cent that the occurrence of 10 or more siginificant parameters is ascribable to chance. Hence the hypothesis of constancy of all parameters of model 63-D has to be rejected.

If the significance of the same ten parameters, this time estimated by the two-stage least-squares method, is looked at (see under B), three of them turn out to be significant on a 5 per cent level, of which one on a 1 per cent level. The above conclusion is not affected by

this outcome in any way, for it should be borne in mind that the changes in the specification of model 63-D have been introduced exclusively on the basis of the estimation results of the ordinary least-squares method, on motives that may be supposed sufficiently known. If nevertheless the isolation of discontinuous changes in the parameters were to be carried out by the two-stage least-squares method, the parameters found would probably not be the same as those pointed out by the method of ordinary least-squares.
The conclusions to be drawn from the results of the test on the constancy of the parameters from model 63-D read as follows.

First. The large number of parameters not deviating significantly from zero, estimated for the observations in the pre- and post-war periods separately, gives rise to serious doubts about the specification of the model chosen by the C.P.B. The specification is not satisfactorily supported by observations from either of the two periods.
Second. Even if the imperfections of the specification of model 63-D are ignored and if one wants to maintain one model for the whole period, it will still be necessary to alter the specification in the sense that binary dummy variables are introduced to account for the changes in some of the parameters.

RESULTS OF THE EXAMINATION OF THE CONSTANCY OF THE PARAMETERS IN MODEL 63-D

For the non-constant parameters a binary dummy variable Z_t has been introduced, which assumes the value 0 in pre-war years and the value 1 in post-war years. Under A are printed the ordinary least-squares estimates, under B the estimates obtained with two-stage least-squares. Non-significant parameters have been underlined.

2. Private gross fixed investments: non-residential

$$\text{A. } I = .65(Z_{-1}-T_z'') + .85c^r_{-1} - .80Z_t c^r_{-1} + \underline{.32}p_i - 17.88\Delta\tilde{w}_1 + \underline{1.76}$$
$$(.14) \quad (.18) \quad (.29) \quad (.26) \quad (2.63) \quad (1.29)$$

post-war coefficient of c^r_{-1}: $+ \underline{.06}$ $(.20)$ $R^2 = .89$ D.W. = 1.91

$$\text{B. } I = .60(Z_{-1}-T_z'') + 1.22c^r_{-1} \underline{-1.14}Z_t c^r_{-1} + \underline{.47}p_i \underline{-14.13}\Delta\tilde{w}_1 + \underline{1.16}$$
$$(.23) \quad (.49) \quad (.64) \quad (.40) \quad (6.62) \quad (2.21)$$

post-war coefficient of c^r_{-1}: $+ \underline{.07}$ $(.48)$

3. Rate of inventory formation

$$\text{A. } N = \underline{.11}v' - .67\tilde{N}_{-1}/\tilde{V}'_{-1} - .52Z_t\tilde{N}_{-1}/\tilde{V}'_{-1} + .15p_m \underline{+.10}K \underline{+1.39}$$
$$(.07) \quad (.14) \quad (.21) \quad (.03) \quad (.19) \quad (.53)$$

post-war coefficient of $\tilde{N}_{-1}/\tilde{V}'_{-1}$: -1.19 $(.22)$

$R^2 = .68$ D.W. = 1.56

$$\text{B. } N = \underline{.22}v' - .52\tilde{N}_{-1}/\tilde{V}'_{-1} \underline{-.41}Z_t\tilde{N}_{-1}/\tilde{V}'_{-1} + .16p_m \underline{-.49}K \underline{+.29}$$
$$(.11) \quad (.16) \quad (.22) \quad (.03) \quad (.49) \quad (.96)$$

post-war coefficient of $\tilde{N}_{-1}/\tilde{V}'_{-1}$: $-.93$ $(.29)$

6. Private employment

A. $$a = .46v_a - .20Z_t v_a + .30K + \underline{.02}c^r_{-1} + .09p_{m-v} - .66\Delta w_{l_{-1}} + \underline{.33}$$

(.06) (.05) (.09) (.02) (.02) (.26) (1.21)

post-war coefficient of v_a: + .27 (.04) $R^2 = .94$ D.W. = 1.76

B. $$a = .46v_a - .20Z_t v_a + \underline{.23}K + \underline{.02}c^r_{-1} + .11p_{m-v'} - \underline{.68}\Delta w_{l_{-1}} + \underline{.30}$$

(.15) (.10) (.29) (.04) (.04) (.51) (.59)

post-war coefficient of v_a: + .26 (.10)

7. Unemployment rate

A. $$\Delta\tilde{w} = -.51a + .33(\Delta\tilde{P} - \Delta\tilde{a}_0)/\tilde{P}_{B-1} - .06\Delta\tilde{T}_c + .06Z_t\Delta\tilde{T}_c + .62\Pi_w - 6.19$$

(.04) (.12) (.02) (.03) (.22) (2.42)

post-war coefficient of $\Delta\tilde{T}_c$: + $\underline{.00}$ (.02) $R^2 = .88$ D.W. = .93

B. $$\Delta\tilde{w} = -.56a + .39(\Delta\tilde{P} - \Delta\tilde{a}_0)/\tilde{P}_{B-1} - .08\Delta\tilde{T}_c + .09Z_t\Delta\tilde{T}_c + .82\Pi_w - 8.25$$

(.05) (.14) (.03) (.04) (.31) (3.25)

post-war coefficient of $\Delta\tilde{T}_c$: + $\underline{.01}$ (.03)

9. Price of consumption

A. $$p_c = -\underline{.02}H_{-\frac{1}{2}} + .59Z_tH_{-\frac{1}{2}} + .24P_{m-\frac{4}{10}} + .50(m-v)_{-\frac{4}{10}} -$$

(.16) (.23) (.04) (.07)

$$-\underline{.04}T'_{k-\frac{1}{3}} + .32Z_tT'_{k-\frac{1}{3}} - .08c^r_{-1} - \underline{.12}$$

(.10) (.11) (.03) (.46)

post-war coefficient of $H_{-\frac{1}{2}}$ resp. $T'_{k-\frac{1}{3}}$: $\underset{(.13)}{+.56}$ and $\underset{(.06)}{+.27}$

$R^2 = .91$ D.W. = 1.36

$$\text{B.}\quad p_c = \underset{(.22)}{\underline{.25}}H_{-\frac{1}{2}} + \underset{(.31)}{\underline{.09}}Z_tH_{-\frac{1}{2}} + \underset{(.04)}{.25}p_{m-\frac{4}{10}} + \underset{(.09)}{.49}(m-v)_{-\frac{4}{10}} +$$

$$+ \underset{(.12)}{\underline{.02}}T'_{k-\frac{1}{3}} + \underset{(.13)}{\underline{.19}}Z_tT'_{k-\frac{1}{3}} - \underset{(.04)}{.12}c^r_{-1} + \underset{(.57)}{\underline{.52}}$$

post-war coefficients of $H_{-\frac{1}{2}}$ resp. $T'_{k-\frac{1}{3}}$: $\underset{(.16)}{+\ .34}$ and $\underset{(.07)}{+\ .21}$

10. Price of investment

$$\text{A.}\quad p_i = \underset{(.23)}{.76}H - \underset{(.25)}{.52}Z_tH + \underset{(.04)}{.38}p_m + \underset{(.08)}{.25}p_{i-1} + \underset{(.45)}{1.49}$$

post-war coefficient of H: $+\ \underset{(.14)}{\underline{.24}}$ $R^2 = .92$ D.W. = 2.40

$$\text{B.}\quad p_i = \underset{(.30)}{\underline{.58}}H - \underset{(.31)}{\underline{.28}}Z_tH + \underset{(.04)}{.39}p_m + \underset{(.09)}{.29}p_{i-1} + \underset{(.49)}{1.21}$$

post-war coefficient of H: $+\ \underset{(.15)}{\underline{.30}}$

11. Price of exports

$$\text{A.}\quad p_b = \underset{(.09)}{.31}H + \underset{(.07)}{.52}p_m + \underset{(.09)}{.36}p'_b - \underset{(.56)}{\underline{1.07}}(\Delta\tilde{w}_l - .48\Delta\tilde{w}_{l-1}) + \underset{(.45)}{\underline{.36}} - \underset{(.73)}{1.99}Z_t$$

post-war constant: $\underset{(.46)}{-\ 1.63}$ $R^2 = .97$ D.W. = 2.48

$$\text{B.}\quad p_b = \underset{(.09)}{.40}H + \underset{(.07)}{.54}p_m + \underset{(.08)}{.29}p'_b - \underset{(0.87)}{3.28}(\Delta\tilde{w}_l - .48\Delta\tilde{w}_{l-1}) + \underset{(.37)}{\underline{.43}} - \underset{(.62)}{2.30}Z_t$$

post-war constant: $- 1.87$

$(.43)$

12. Price of autonomous expenditure

A. $p_x = .86H - .45Z_tH + \underline{.05}\Delta H + .27p_m + .16Z_tp_m + .26p_{x-1} + 1.17$

$(.21)(.21)\quad(.12)\quad(.04)\quad(.08)\quad(.08)\quad(.34)$

post-war coefficients of H resp. p_m: $+ .41$ and $+ .43$

$(.13)\quad(.07)$

$R^2 = .95 \quad D.W. = 1.96$

B. $p_x = .73H - \underline{.33}Z_tH + \underline{.22}\Delta H + .25p_m + \underline{.17}Z_tp_m + .32p_{x-1} + .99$

$(.33)(.31)\quad(.24)\quad(.05)\quad(.09)\quad(.12)\quad(.43)$

post-war coefficients of H resp. p_m: $+ .42$ and $+ .40$

$(.07)\quad(.17)$

5.3. SOME CALCULATIONS RELATING TO THE POWER OF CHOW'S TEST

In section 4.3 the theoretical derivation of the power of Chow's test has been given. It may be interesting to mention here a few results of the new method. To that end, for each equation of model 63-D three calculations of the power of Chow's test will be carried out. First, the drawing-up of alternative hypothesis will be discussed.

The null hypothesis of Chow's test is that for each equation the parameters of the pre-war period are equal to those of the post-war period. In (4.1.5) this null hypothesis is represented by $\underline{\beta}_1 = \underline{\beta}_2 = \underline{\beta}_0$; in the present example $\underline{\beta}_1$ represents the pre-war parameter of a reaction

equation, and $\underline{\beta}_2$ the post-war corresponding parameters. As appears from (4.2.6) and (4.2.9), this null hypothesis can also be denoted by: $\underline{\gamma} = \underline{\beta}_2 - \underline{\beta}_1 = 0$. As an alternative hypothesis for $\underline{\gamma}$ any vector of numbers can be chosen. An alternative hypothesis for $\underline{\gamma}$ can also be arrived at, however, by filling in for the elements of $\underline{\beta}_2$ and $\underline{\beta}_1$ values that are 'not unreasonable', and calculating the difference between them. This procedure will be followed. For the estimated parameters in $\hat{\underline{\beta}}_1$ and $\hat{\underline{\beta}}_2$ the corresponding standard deviations emerge from the procedure described on p. 167. If as 'not unreasonable' values the 2s-limits of a coefficient from $\hat{\underline{\beta}}_2$ and of the corresponding coefficient from $\hat{\underline{\beta}}_1$ are chosen, the maximum (absolute) magnitude of the difference between the two coefficients can be calculated; this value (with its relevant sign) will serve as one of the alternative hypothesis for the element of $\underline{\gamma}$ concerned. If, for example, one has a set of corresponding elements from $\hat{\underline{\beta}}_2$ and $\hat{\underline{\beta}}_1$ with respective values of 0.30 and 0.40, and with respective standard deviations of 0.05 and 0.10, then the 2s-limits lie at, respectively, 0.20 and 0.40, and 0.20 and 0.60. The maximum difference between the two coefficients will then be 0.20 - 0.60 = - 0.40; that number will be chosen as alternative hypothesis for the relevant element of $\underline{\gamma}$. Similarly three alternative hypothesis have been developed for all 13 reaction equations, for the 1s- and 3s-limits as well as for the 2s-limits. The results obtained are to be found in the tables below.
To appreciate the results of both tables account must be taken of the relatively large standard errors of the pre- and post-war coefficients of the model. Alternative hypothesis developed with the help of these standard errors will imply values of the differences that are sometimes very high. As a consequence, the unreliability

Table 5.3a. Power of Chow's test, calculated for the parameters from the reaction equations of model 63-D. The significance level is 5%.

Alternative hypothesis*	Equations												
	1	2	3	4	5	6	7	8	9	10	11	12	13
1s-limits	.70	.58	.37	.60	1.00	.93	.56	.09	1.00	.24	.70	.66	.31
2s-limits	1.00	.98	.60	.95	1.00	1.00	1.00	.12	1.00	.36	.97	.98	.81
3s-limits	1.00	1.00	.81	1.00	1.00	1.00	1.00	.16	1.00	.50	1.00	1.00	.99

Table 5.3b. Power of Chow's test, calculated for the parameters from the reaction equations of model 63-D. The significance level is 1%.

Alternative hyopthesis*	Equations												
	1	2	3	4	5	6	7	8	9	10	11	12	13
1s-limits	.42	.32	.16	.33	1.00	.76	.30	.02	.99	.08	.43	.39	.12
2s-limits	1.00	.89	.33	.82	1.00	.98	1.00	.03	1.00	.15	.87	.88	.55
3s-limits	1.00	1.00	.57	.99	1.00	1.00	1.00	.05	1.00	.25	.99	1.00	.94

* For the meaning of the expressions 1s-, 2s- and 3s-limits, see the text of this section.

of the pre- and post-war coefficients tends to flatter the picture presented by the calculations about the power of Chow's test. Nevertheless the tables 5.3a and 5.3b point out that the power remains in most cases far below 1; the value 1 indicates in this connection the probability that the alternative hypothesis is rightly accepted. With the 2s-limits the picture is much more favourable than with the 1s-limits. The alternative hypothesis have by then climbed to relatively very high values of the differences, however. From the above results it must be concluded that owing to the unreliability of the estimates of the coefficients in model 63-D, which had already been indicated earlier, no representative outcome is achieved as to to the power of Chow's test. Further research with respect to other models is certainly justified, however.

APPENDIX TO CHAPTER V. THE C.P.B.-MODEL 63-D

The following estimated parameters of the model have been carried out with:

A. the ordinary least-squares method, and
B. the two-stage least-squares method.

Observations from the years 1923 through 1938 and 1949 through 1965 have been used, 33 years in total, 16 pre-war and 17 post-war. The parameters have been supposed constant, in accordance with the attitude of the C.P.B. The standard deviations are given (between brackets) below the relevant coefficients. The parameters not significant on a 5 per cent level have been underlined. For the ordinary least-squares estimates, the corresponding values of R^2 and of the Durbin-Watson test statistic have also been given; with the help of this test statistic a statement can be made[13] about the presence of absence of autocorrelation in the disturbances.
Model 63-D, in the specification of the C.P.B. and with the coefficients calculated by TR-4 of the State University of Groningen, reads as follows: (For the meaning of the symbols, see pp. 185, 186, and 187).

---

13. A table of the Durbin-Watson test statistic is given in: C.F. Christ, Econometric models and methods, New York 1966, p. 672. None of the DW-values mentioned reaches a significant value on a level of 5 per cent, which means that the assumption of absence of autocorrelation in the disturbances can be maintained.

REACTION EQUATIONS

I. Expenditure categories

1. Total private consumption

A. $C = .85L^{B}_{-\frac{1}{4}} + \underline{.07}Z^{B}_{-\frac{3}{4}} + .30\Delta p_{c} + .08c^{r}_{-1} - .21\Delta C_{-1} - 1.08$

$\quad\quad (.06) \quad (.04) \quad (.08) \quad (.03) \quad (.06) \quad (.37)$

$R^2 = .94 \qquad D.W. = 1.45$

B. $C = .77L^{B}_{-\frac{1}{4}} + .14Z^{B}_{-\frac{3}{4}} + .27\Delta p_{c} + \underline{.07}c^{r}_{-1} - .23\Delta C_{-1} - .99$

$\quad\quad (.07) \quad (.05) \quad (.09) \quad (.04) \quad (.06) \quad (.37)$

2. Private gross investments: non-residential

A. $I = .63(Z_{-1} - T''_{z}) + .51c^{r}_{-1} + .71p_{i} - 14.73\Delta\tilde{w}_{1} + \underline{1.44}$

$\quad\quad (.15) \quad (.14) \quad (.24) \quad (2.65) \quad (1.44)$

$R^2 = .86 \qquad D.W. = 1.75$

B. $I = .66(Z_{-1} - T''_{z}) + .63c^{r}_{-1} + .65p_{i} - 12.94\Delta\tilde{w}_{1} + \underline{1.07}$

$\quad\quad (.15) \quad (.23) \quad (.25) \quad (.21) \quad (1.42)$

3. Rate of inventory formation[14]

A. $N = \underline{.12}v' - .75\tilde{N}_{-1}/\tilde{V}'_{-1} + .14p_{m} - \underline{.05}K + \underline{.89}$

$\quad\quad (.08) \quad (.15) \quad (.03) \quad (.20) \quad (.54)$

$R^2 = .60 \qquad D.W. = 1.38$

---

14. In the C.P.B.-version the variable t^{*} occurs as an additional regressor. This variable equals zero for every post-war year. Consequently the coefficient would be indefinite for the post-war period only; for this reason it was decided to omit this variable.

B. $N = .28v' - .46\tilde{N}_{-1}/\tilde{V}'_{-1} + .16p_m - 1.01K - \underline{.74}$

(.11) (.18) (.03) (.46) (.92)

4. Exports of commodities

A. $b = 1.36b_c - 1.20(p_b - p_b') - .69(p_b - p_b')_{-1} +$

(.12) (.27) (.26)

$+ 5.17\Delta\tilde{w}_{1-\frac{5}{12}} - .49\Delta p_{v'} + \underline{.31}$

(1.92) (.18) (.83)

$R^2 = .91$ D.W. = 2.38

B. $b = 1.41b_c - 1.28(p_b - p_b') - .53(p_b - p_b')_{-1} +$

(.14) (.52) (.26)

$+ 7.20\Delta\tilde{w}_{1-\frac{5}{12}} \underline{- .40}\Delta p_{v'} + \underline{.10}$

(3.36) (.23) (.83)

II. Factors and capacity

5. Imports of commodities[15]

A. $m = 1.46v_m \underline{+ .09}\Delta v_m + 2.21N - .27p_{m-v'} \underline{+ .20}\Delta p_v' \underline{- .31}\Delta\tilde{w}_{1-\frac{1}{2}} - 2.07$

(.14) (.12) (.37) (.11) (.23) (1.63) (.97)

$R^2 = .91$ D.W. = 2.63

---

15. The C.P.B. version contains in this equation also the variables k and k'. As these variables consists only of zeros in one of the two periods, they are an inpediment to estimating parameters for individual periods. For that reason they have been omitted.

B. $m = 1.57v_m + \underline{.16}\Delta v_m + 3.63N - \underline{.21}p_{m-v'} - \underline{.51}\Delta p'_v - \underline{.78}\Delta\tilde{w}_{l-\frac{1}{2}} - \underline{2.24}$

(.29) (.35) (1.35) (.26) (.81) (3.86) (1.92)

6. Private employment

A. $a = .27v_a + .43K + .05c^r_{-1} + .10p_{m-v'} - .93\Delta\tilde{w}_{l-1} + \underline{.50}$

(.05) (.11) (.02) (.02) (.32) (.27)

$R^2 = .90$ D.W. = 1.38

B. $a = .23v_a + .43K + .07c^r_{-1} + .11p_{m-v'} - 1.01\Delta\tilde{w}_{l-1} + \underline{.66}$

(.06) (.17) (.03) (.02) (.31) (.35)

7. Unemployment rate

A. $\Delta\tilde{w} = -\ .52a + .31(\Delta\tilde{P} - \Delta\tilde{a}_0)/\tilde{P}_{B-1} - \underline{.03}\Delta\tilde{T}_c + .60\Pi_w - 5.84$

(.04) (.12) (.02) (.24) (2.54)

$R^2 = .86$ D.W. = 1.22

B. $\Delta\tilde{w} = -\ .55a + .35(\Delta\tilde{P} - \Delta\tilde{a}_0)/\tilde{P}_{B-1} - \underline{.03}\Delta\tilde{T}_c + .78\Pi_w - 7.73$

(.05) (.12) (.02) (.24) (2.62)

III. Wages and prices

8. Wage rate

A. $l = .46p_c + .18\{(v'-a)_{-1} - .68(v'-a)_{-2}\} -$

(.10) (.15)

$- \underline{1.06}(\Delta\tilde{w}_l - .68\Delta\tilde{w}_{l-1}) - 1.94(\Delta\tilde{w}_{l-1} - .68\Delta\tilde{w}_{l-2}) + .66l_{-1} + \underline{.67}$

(.73) (.75) (.08) (.41)

$R^2 = .89$ D.W. = 1.67

B. $l = .37p_c + \underline{.06}\{(v'-a)_{-1} - .68(v'-a)_{-2}\} -$

(.14) (.17)

$- 3.30(\Delta\tilde{w}_l - .68\Delta\tilde{w}_{l-1}) - 2.82(\Delta\tilde{w}_{l-1} - .68\Delta\tilde{w}_{l-2}) + .74l_{-1} + \underline{.59}$

(1.52) (.95) (.10) (.43)

9. Price of consumption

A. $p_c = .37H_{-\frac{1}{2}} + .28p_{m-\frac{4}{10}} + .53(m-v)_{-\frac{4}{10}} + .18T'_{k-\frac{1}{3}} - .10c^r_{-1} + \underline{.20}$

(.10) (.04) (.08) (.06) (.03) (.46)

$R^2 = .87$ D.W. = 1.56

B. $p_c = .36H_{-\frac{1}{2}} + .28p_{m-\frac{4}{10}} + .55(m-v)_{-\frac{4}{10}} + .17T'_{k-\frac{1}{3}} - .12c^r_{-1} + \underline{.25}$

(.10) (.04) (.07) (.06) (.03) (.34)

10. Price of investments

A. $p_i = .37H + .41p_m + .33p_{i-1} + .89$

(.13) (.04) (.07) (.37)

$R^2 = .90$ D.W. = 2.51

B. $p_i = .35H + .42p_m + .34p_{i-1} + .89$

(.15) (.04) (.07) (.35)

11. Price of exports

A. $p_b = \underline{.15}H + .58p_m + .28p_b' - \underline{.86}(\Delta\tilde{w}_l - .48\Delta\tilde{w}_{l-1}) - .61$

(.08) (.08) (.10) (.62) (.30)

$R^2 = .97$ D.W. = 2.17

B. $p_b = \underline{.19}H+.62p_m+\underline{.21}p_b'-2.93(\Delta\tilde{w}_l-.48\Delta\tilde{w}_{l-1})-.67$

(.10) (.09) (.12) (1.31) (.33)

12. Price of autonomous expenditure

A. $p_x = .48H+\underline{.19}\Delta H+.32p_m+.36p_{x-1}+.76$

(.13) (.11) (.03) (.07) (.28)

$R^2 = .94$ D.W. = 2.03

B. $p_x = .39H+.45\Delta H+.31p_m+.42p_{x-1}+.71$

(.16) (.20) (.04) (.09) (.29)

IV. Liquidities

13. Deposits

A. $c^r_{-1} = \underline{.41}\tilde{E}+1.37\tilde{E}_{-1}+.32V'-6.59r_{k-1}+1.19N_{-2}+1.05r_e +$

(.36) (.51) (.15) (1.20) (.44) (.36)

$+ \underline{.07}p'_{v-1}\underline{-.61}$

(.26) (1.26)

$R^2 = .84$ D.W. = 2.58

B. $c^r_{-1} = \underline{.74}\tilde{E}+1.26\tilde{E}_{-1}+.31V'-6.66r_{k-1}+1.12N_{-2}+.96r_e +$

(.81) (.39) (.15) (1.06) (.39) (.37)

$+ \underline{.04}p'_{v-1}\underline{-.81}$

(.28) (1.44)

DEFINITIONAL EQUATIONS

I Value-volume relationships

14. $C = c + p_c$
15. $X = x + p_x$
16. $I = i + p_i$
17. $B = b + p_b$
18. $M = m + p_m$
19. $V' = v' + p_v'$
20. $V = v + p_v'$

II Expenditure totals

21. $v' = .48c + .14x + .11i + .27b$
22. $v_a = .46c + .20x + .16i + .18b$
23. $v_m = .45c + .15x + .11i + .29b$
24. $p_v' = .47p_c + .14p_x + .11p_i + .27p_b$
25. $V = .44C + .14X + .10I + .25B + .92N + .06D$

III Balance of payments

26. $\tilde{E} = .27B - .35M + .07D + .93\tilde{E}_{-1}$

IV Costs and margins

27. $H = 1 - (v'-a)_{-\frac{1}{2}}$
28. $p_{m-v'} = p_m - p_{v'-\frac{1}{2}} + .06T'_{k-\frac{1}{3}}$
29. $K = p_v' - .27l - .30p_m - .06T'_{k-\frac{1}{3}}$

V Unemployment

30. $\tilde{w} = \Delta\tilde{w} + \tilde{w}_{-1}$

31a. $\Delta\tilde{w}_1 = 4.34\Delta \ln(\tilde{w}+2) - .20\Delta\tilde{w}$

31b. $\Delta\tilde{w}_1 = \{4.34/(\tilde{w}_{-1}+2) - .20\}\Delta\tilde{w}$

32. $\Pi_w = \Pi\{\tilde{P}_{-1}/\tilde{P}_{B-1}\}$ — $\Pi_w = 4.03$ for the period 1923-1938; $\Pi_w = 4.66$ for the period 1949-1965

VI Incomes

33. $L = a + l$

34. $Z = 3.77V - 1.06L - .24T_k - 1.23M - .24F$

35. $L^B = .87L + .87O'_L$

36. $Z^B = 1.50Z + 1.50O'_Z$

VII Taxes

37. $T_k = V' + T'_k$

38. $\tilde{T}''_z = \Delta(\tilde{T}_z/\tilde{Z}_{-1})$

List of symbols

(Symbols without special indication refer to relative changes. Absolute quantities are indicated by $\sim$. Capitals refer to values, lower-case type to volumes and prices).

	a	Number of persons employed in enterprises* (man-years)
	a_0	Number of persons employed in goverment sector (man-years)
B	b	Exports of commodities
	b_c	Competing exports
C	c	Total private consumption
	c^r_{-1}	Time deposits plus demand deposits at the end of the year
D		Net invisible exports
$\tilde{E}$		Balance of payments as a percentage of V' (total expenditure less inventory changes and net invisibles)
F		Depreciation
H		Labour costs per unit (equation 27)
I	i	Private gross fixed investment, excluding dwellings
	k	Quantitative import restrictions (1932-1937)
	k'	Measure of rate of liberalization (1949-1955)
K		Gross profits per unit of output (equation 29)

* 'Enterprises', in the usage of this model, include public utilities and government enterprises 'industry' is also used in this sense.

	l	Average gross wages per standard year of 300 days
L		Wage-bill of enterprises
L^B		Disposable labour income
M	m	Imports of commodities
N		Inventory changes, their rate of change being expressed as a percentage of lagged total expenditure less inventory changes and net invisibles
O'_L		Net income transfers with respect to labour income, including government wages and direct taxes based on cash receipts
O'_z		Net income transfers with respect to non-labour income, including direct taxes based on cash receipts
P		Population of working age (14-65 years)
P_B		Dependent working population
	p_b	Price of exports
	p'_b	Price of competing exports
	p_c	Price of consumption
	p_i	Price of investment
	p_m	Price of imports
	$p_{m-v'}$	Margin between price of imports adjusted for indirect taxes and price of total expenditure less inventory changes and net invisibles (see equation 28)
	p'_v	Price of total expenditure less inventory changes and net invisibles
	p_x	Price of autonomous expenditure
Π_w		See equation 32
	r_e	Exchange rate
	r_k	Short-term rate of interest
$\tilde{T}_c$		Minimum temperature below 0^0 Centigrade (sum of monthly averages)

$\tilde{T}_k$		Indirect taxes minus subsidies (amount)
$\tilde{T}'_k$		Incidence of indirect taxes minus subsidies ($\tilde{T}'_k = \tilde{T}_k/\tilde{V}'$)
	t^*	Pre-war decreasing trend (1923 = 15; 1938 = 0)
$\tilde{T}_z$		Direct taxes paid by non-wage earners on basis of cash receipts (amount)
T''_z		Variation in the incidence of direct taxes on non-labour income (see equation 38)
V	v	Total expenditure
V'	v'	Total expenditure less inventory changes and net invisibles
	v_a	Total expenditure less inventory changes and net invisibles, components reweighted by intensities of labour demand (see equation 22)
	v_m	Total expenditure less inventory changes and net invisibles, components reweighted by import requirements (see equation 23)
	$\tilde{w}$	Unemployment as percentage of working population
	$\tilde{w}_1$	Curvilinear indicator of use of capacity (see equation 31a or 31b)
X	x	Autonomous expenditure (government expenditure, investment of government enterprises and construction of dwellings)
Z	z	Non-labour income
Z^B		Disposable non-labour income

Bibliography

Abrahamse, A.P.J. and R.A.I. van Frederikszust, Discriminant analysis and the prediction of corporate failure, Working Paper R/75/20, Graduate School of Management, Delft, August 1975.

Amundsen, H.T., Binary variable multiple regressions, Scandinavian Journal of Statistics, Vol. 1 (1974), pp. 59-70.

Bartlett, M.S., Fitting a straight line when both variables are subject to error, Biometrics, Vol. 5 (1949), pp. 207-212.

Blumenthal, T., A test of the Klein-Shinkai econometric model of Japan, International Economic Review, Vol. 6 (1965), no. 2, pp. 211-228.

Central Bureau of Statistics (C.B.S.), Comparison of cost and productivity analysis by means of grouping with econometric analysis, Statistische en Econometrische Onderzoekingen, First and second quarter, 1963, pp. 43-47.

Chow, G.C., Tests of equality between sets of coefficients in two linear regressions, Econometrica, Vol. 28 (1960), no. 3, pp. 591-605.

Christ, C.F., Econometric models and methods, New York 1966.

Clelland, R.C., J.S. de Cani, F.E. Brown, J.P. Bursk and D.S. Murray, Basic statistics with business applications, New York 1966.

Cochran, W.G., Analysis of covariance: its nature and uses, Biometrics, Vol. 13, September

1957, pp. 261-281.

Desai, M., An econometric model of the world tin economy, Econometrica, Vol. 34 (1966), no. 1, pp. 105-134.

Dixon, W.J. and F.J. Massey, Introduction to statistical analysis, New York 1957.

Duesenberry, J.S., G. Fromm, L.R. Klein and E. Kuh (Ed.), The Brookings quarterly econometric model of the United States, Amsterdam 1965.

Duesenberry, J.S., G. Fromm, L.R. Klein and E. Kuh (Ed.), The Brookings model: some further results, Amsterdam 1969.

Fisher, F.M., The choice of instrumental variables in estimation of economic-wide economic models, International Economic Review, Vol. 6 (1965), no. 3, pp. 245-274.

Fisher, F.M., Tests of equality between sets of coefficients in two linear regressions: an expository note, Econometrica, Vol. 38 (1970), no. 2, pp. 361-366.

Fisher, J.A., An analysis of consumer goods expenditures in 1957, Review of Economics and Statistics, Vol. 44 (1962), pp. 64-71.

Fourgeaud, C., L'estimation dans les modèles a composante saisonnière, Thesis defended at the University of Paris, 1955.

Gibson, W.M. and G.H. Jowett, 'Three-group' regression analysis, part I and II, Applied Statistics, Vol. 6, (1957), pp. 114-122 and 189-197.

Goldberger, A.S., Econometric theory, New York 1964.

Graybill, F.A., An introduction to linear statistical models, Vol. 1, New York 1961.

Gujarati, D., Use of dummy variables in testing for equality between sets of coefficients in two linear regressions: a note, The American Statistician, Vol. 24, no. 1, pp. 50-52.

Hazlewood, A. and P. Vandome, A post mortem on econometric forecasts for 1959, Oxford University Institute of Statistics Bulletin, Vol. 23, February 1961, pp. 67-81.

Henshaw, R.C., Application of the general linear model to seasonal adjustment of economic time series, Econometrica, Vol. 34 (1966), no. 2, pp. 381-395.

Hervey, J.L., A regional analysis of the effects of age, education, and occupation on median income, Journal of Regional Science, Vol. 6 (1966), no. 2, pp. 35-48.

Hicks, C.R., Fundamental concepts in the design of experiments, New York 1963.

Hooper, J.W. and H. Theil, The extension to Wald's method of fitting straight lines to multiple regression, Revue de L'Institute International de Statistique, Vol. 26 (1958), pp. 37-47.

Huang, D.S., Regression and econometric methods, New York 1970.

Hudson, D.J., Fitting segmented curves whose join points have to be estimated, Journal of the American Statistical Association, Vol. 61, December 1966, pp. 1097-1129.

Ichimura, S., L.R. Klein, S. Koizuma, K. Sato and Y. Shinkai, A quarterly econometric model of Japan, 1952-1959, Osaka Economic Papers,

Vol. 12 (1964).

Janosi, P.E. de, Factors influencing the demand for new automobiles: a cross-section analysis, Thesis defended at the University of Michigan 1956.

Johnson, T., Qualitative and limited dependent variables in economic relationships, Econometrica, Vol. 40 (1972), no. 3, pp. 455-462.

Johnston, J., Econometric methods, 2nd edition, New York 1972.

Kendall, M.G. and W.R. Buckland, Dictionary of statistical terms, London 1960.

Kendall, M.G. and A. Stuart, The advanced theory of statistics, Vol. 1 and Vol. 3, London 1958, respectively 1966.

Klein, L.R., R.J. Ball, A. Hazlewood and P. Vandome, An econometric model of the United Kingdom, Oxford 1961.

Klein, L.R., A. Hazlewood and P. Vandome, Re-estimation of the econometric model of the United Kingdom and forecasts for 1961, Oxford University Institute of Statistics Bulletin, Vol. 23, February 1961, pp. 49-66.

Klein, L.R. and Y. Shinkai, An econometric model of Japan, 1930-1959, International Economic Review, Vol. 4 (1963), no. 1, pp. 1-28.

Kloek, T., Note on convenient matrix notations in multivariate statistical analysis and in the theory of linear regression, International Economic Review, Vol. 2 (1961), no. 3, pp. 351-360.

Kloek, T. and L.B.M. Mennes, Simultaneous equations estimation based on principal components of pre-determined variables, Econometrica, Vol. 28 (1960), no. 1, pp. 45-61.

Kooyman, M.A., The use of dummy variables in linear relations, paper presented at the European Econometric Society Meeting, Budapest 1972.

Kullback S. and H.M. Rosenblatt, On the analysis of multiple regression in k categories, Biometrica, Vol. 44 (1957), pp. 67-83.

Ladd, G.W., Linear probability functions and discriminant functions, Econometrica, Vol. 34 (1966), no. 4, pp. 873-885.

Lee, T.H., Demand for housing: a cross-section analysis, Review of Economics and Statistics, Vol. 45 (1963), pp. 190-196.

Lehman, L.H., Testing statistical hypothesis, New York 1959.

Lüdeke, D., Ein Ökonometrisches Vierteljahresmodel für die Bundesrepublik Deutschland, Tübingen 1969.

Malinvaud, E., Statistical methods of econometrics, Amsterdam 1966.

Menges, G., Ökonometrische Prognosen, Cologne und Opladen 1967.

Menges, G. and H. Diehl, Time stability of structural parameters, Colston Papers no, 16, London 1964, pp. 299-320.

Nerlove, M., Two models of the British economy: a fragment of a critical study, International Economic Review, Vol. 6 (1965), no. 2, pp. 127-181.

Neter, J. and E.S. Maynes, On the appropriateness of the correlation coefficients with a 0,1 dependent variable, Proceedings of the Business and Economic Statistics Section of the American Statistical Association 1967, pp. 447-453.

Orcutt, G.H., M. Greenberger, J. Korbel and A.M. Rivlin, Microanalysis of socioeconomic systems: a simulation study, New York 1961.

Roscoe, J.T., Fundamental research statistics for the behavioral sciences, New York 1969.

Schilderinck, J.H.F., Een ekonometrisch model van de Nederlandse ekonomie, Thesis defended at the Catholic University at Tilburg, 1970.

Snedecor, G.W. and W.G. Cochran, Statistical methods, Ames Iowa 1967.

Suits, D.B., Use of dummy variables in regression equations, Journal of the American Statistical Association, Vol. 52, December 1957, pp. 548-551.

Tate, R.F., Applications of correlation models for biserial data, Journal of the American Statistical Association, Vol. 50, December 1955, pp. 1078-1095.

Theil, H., Economic forecasts and policy, Amsterdam 1958.

Theil, H., Principles of econometrics, New York 1971.

Thomas, J.J. and K.F. Wallis, Seasonal variation in regression analysis, Journal of the Royal Statistical Society, Series A, Vol. 134 (1971), pp. 57-73.

Index